COAST, ESTUARY and SEASHORE LIFE

COAST, ESTUARY and SEASHORE LIFE

Alan P. Major
Text Illustrations by
Barbara Prescott

JOHN GIFFORD
LONDON

First Published in Great Britain 1973
by John Gifford Ltd, 125 Charing Cross Road,
London WC2H OEB

ISBN 07077 0003 8

Dedicated to
my daughters
STEPHANIE AND CAROLINE
that they may always be
able to discover and be
fascinated by the wonders
of the seashore

Text set Baskerville eleven on twelve point and printed in
Great Britain by Tonbridge Printers.
Colour printed by Tindall Press, Chelmsford.
Books bound by Mackays, Chatham.

CONTENTS

Acknowledgments

The author wishes to record, with great pleasure, his sincere thanks however insufficient, for the kind assistance of the following during th preparation of this book: to Sir Maurice Yonge, the eminent marin biologist, for his encouragement and allowing me to use a shore zona system similar in descriptive terms to that used in his own books; t Mr. George H. Haines for his excellent photographs of coastline anc shore scenery; to the Royal Society for the Protection of Birds and th British Trust for Ornithology for advice; also equally important, to m publisher for his patience.

I gladly record my thanks to Miss Barbara Prescott, for her advice co-operation and generous assistance in the skilled creation of her tex illustrations, which made my task in writing this book much easier.

Lastly, but by no means least, sincere thanks to my wife for checkin the manuscript.

Chapter 1

Description of Habitats

The British Isles are fortunate in having a very varied coastline, anging from sandy bays, salt marshes and mudflats, boulder-strewn hores, to restless shingle, constantly eroding chalk cliffs and almost ermanent granite, etc. This in turn means that the flora and fauna re also richly numerous, yet can differ greatly within the space of a ew miles. One type of habitat may contain a prolific total of species, hereas another will be almost barren for one or several reasons, as I ave indicated in the relevant description.

The high-water mark is the limit reached by the tide on the coast or n the estuary. This may be indicated by a line of refuse, seaweed, dead reatures and other debris lying on the shore and left behind by the eceding water. The shape of the shore may also indicate high-water ıark, the latter's extent ending where the sloping shingle washed by ıe sea ends and a level platform of dry shingle, pebbles, etc., begins. third high-water level indicator along the length of cliffs may be a ow of lichens – small green, yellow or black moss-like plants – below hich the cliff is discoloured by the salt of the seawater, the cliff surface ometimes also being worn smooth by the sea's action. Seaweeds may be rowing on the cliff face below this line, which is also often riddled by ıe boring of marine creatures, or bears various shellfish, Limpets, etc. bove the high-water line various flowering plants can grow on the iff face.

It should be mentioned that there is no visible low-water mark in the me sense; only an area, the shore or inter-tidal zone, between low-ater and high-water. The flora and fauna of this shore or inter-tidal ne of our coasts and estuaries, with the flora and fauna of above gh-water level, are the subject of this book.

For the purpose of aiding identification of species the shore or inter-dal zone is divided into three sections – the upper, middle and lower

Fig. 1 The Divisions of the Shore Zones

shore (Fig. 1). Thus if a species is stated to frequent the lower shore the reader will know exactly to what area of the shore he must go to in order to find that species. The upper shore is basically the area of the shore which is not usually covered by water unless during Spring tides. This part of the shore above high water level has less animal species than the other two, but, if suitable, is the area most favoured by human beings, who sometimes lie on it like sardines in a tin. Beyond the upper shore, on the inland side, there is sometimes also a splash zone, particularly where cliffs and high coastline stand close behind the shore. As its name indicates, it is splashed with spray from the pounding surf at high tide and in this zone some plant and animal life exists. The middle shore is the area from average high water level to average low water level and every day is twice uncovered to the air and twice submerged by the sea. Despite this, the middle shore contains a great many plant and animal species, and human beings also take it over for their use when the tide is out. The lower shore is the area descending seawards from the average low water level down to the shore limit at the low water level of the Spring tides. It is usually covered by water, except during the Spring tides. It contains an abundance of animal and plant life; frequently also marine life from the shallow sub-littoral zone beyond and the deeper sea bottom.

Space does not permit minute detailed description of the many varying types of coast, estuary and seashore and their even more numerous combinations. The following descriptions, however, cover all the basic details needed to identify the types of habitation and thus the reader can gauge what flora and fauna can be expected to be found in that situation.

CLIFFS:

BASALT CLIFFS

Basalt is a fine-grained, crystalline, dark reddish-brown rock, containing the minerals augite and feldspar. It is either columnar, tabular or globular in structure. The columnar type is straight or curved, sometimes almost horizontal, the forms of the columns being either octagonal, hexagonal or pentagonal. One example of this is the rock pillars of the Giant's Causeway, Northern Ireland, which although exposed to the elements for centuries have angles still perfectly and closely fitting together. The isle of Staffa, with Fingle's Cave, is another remarkable example of basalt columns. Basalt also forms the majority of other Scottish western isles, such as Skye, Muck, Mull and Eigg and a part of the Argyll mainland, where it occurs as sheer, almost vertical, cliffs. When softer rocks are eroded the basalt often remains as capes and islands, examples being the Farne islands and the coast at Bamburgh, Northumberland. Basalt also often assumes curious shapes and rounded masses.

CHALK CLIFFS AND SHORE PLATFORMS

Chalk is a pure, soft limestone, opaque white in colour. Usually formed by the accumulation of the shells of Foraminifera – minute marine creatures – and marine sediment with larger marine organisms. The Upper Chalk level may contain Flints, and Fossils of Sea Urchins, Corals, etc., evidence of when the chalk was a marine deposit, are often found within the chalk. The chalk usually forms sheer, steep, vertical or almost vertical cliffs, situated chiefly on the eastern and southern coasts of England, including the Isle of Wight, particularly Kent, Sussex, Dorset, Lincolnshire and Yorkshire. Occasionally, through pressure, the chalk becomes harder and more typically rock-like and resistant to the sea's penetration. It is eroded by the sea and rain, which penetrates cracks and fissures. During winter this water content freezes then expands, causing the chalk to become unstable, disintegrate and eventually collapse. Where this occurs the upper foreshore immediately below may be littered with various-sized boulders of chalk worn smooth by the sea, or be covered with a layer of soft chalk made spongy by every incoming tide, and the sea is also white as milk while it covers the fallen chalk. Where the sea has eroded the chalk cliff, so that the latter has retreated through centuries of attack, the foreshore may have a semi-hard, slippery platform base of chalk, worn smooth and containing varying depth depressions as pools at low water. It is also littered with lumps of chalk covered with seaweeds, various shellfish and other marine creatures. The sea's action creates caves in the chalk cliffs, deep gullies and chasms, buttresses or isolated chalk pinnacles and offshore reefs. The top of the chalk cliff extending inland usually bears the typical short, chalk-Downland turf species of grass and flowers with relevant insects.

GRANITE CLIFFS

Granite is a common, coarse-grained, crystalline, igneous rock, containing various minerals, including quartz, mica, hornblende, feldspar. It was formed by pressure deep in the earth but varies considerably in resistance and hardness and has fissures and joints which are places of weakness. At these points the sea and weather erode away the less solid rock to form a large mass, blocks, walls and buttresses, giving it a castellated appearance. Or the sea may also create caves and coves from hollows. Where they occur, due to their coloration, either steep or sloping, the granite cliffs form a beautiful part of the coastal scenery. Granite varies in colour, usually light, depending on the mineral content, pink and grey, red, whitish-grey, bluish-grey. The top of the cliffs and sometimes the ledges below it support some of the typical coastal vegetation. Granite is found chiefly on the Cornish coast, particularly Land's End, the Scilly Isles – which are entirely composed of granite – Peterhead, on the east coast of Scotland, and the Isle of Rum. At Peterhead the red hue of the granite is due to its feldspar being flesh-coloured.

LIMESTONE CLIFFS

Limestone is a sedimentary rock and owes its origin to the depositing of remains of sea organisms on the old ocean beds. Often fossil marine shells and skeletons of ancient sea creatures are found embedded in it. Most of the varieties of limestone contain silica, alumina, calcium-magnesium carbonate and iron compounds. It is soft, usually pale-coloured, and can be easily scratched, this also means it is readily eroded by the sea's action upon its surface or by penetration of its joints. The latter creates caves, inlets, cauldrons and blow-holes in the sheer or sloping cliffs, which may also have numerous stacks or reefs in close proximity. Chalk is a variety of limestone and so is oolite, a white or yellow granular rock. Some examples of limestone coast are the Great Orme, North Wales, the Tenby Peninsular, Wales, the Durham coast and Durness, Scotland. The top of the limestone cliff frequently supports the characteristic Downland flora similar to the chalk cliffs.

SAND AND CLAY CLIFFS

Gault is the geological term for several stiff, sometimes calcareous, sandy clays, which vary in colour from pale grey to blue, examples of which occasionally occur on the coast as cliffs. For example, there is a good example of gault at Folkestone, Kent. Clay cliffs, sometimes with large and small stones and flint, can be notoriously unstable and are constantly collapsing or sliding towards the sea, usually because of rain or seeping water (which make them become slippery), hot weather, which causes them to dry out and split into widening cracks, or erosion of the underlying layer by the sea.

Cliffs of sand and gravel, called crags, are soft and loose, and are readily eroded by the sea so that they collapse. This type of cliff occurs along much of the Norfolk and Suffolk coast. On the more stable areas a sand and clay loving flora is supported.

SANDSTONE CLIFFS

Sandstone consists of sand-sized grains of quartz, compactly cemented to form a hard, coarse rock, and may also contain feldspar, mica, clay and pebbles, to form a conglomerate or matrix of materials. It varies in colour from greenish-grey to the more familiar red-brown, but is sometimes variegated or has a striped appearance. It occurs in massive cliff beds or it forms typical flagstones. Although an extremely hard rock, which is eroded very slowly, it does have dividing bedding-planes and joints which are penetrated by the sea. Where this occurs caves, blow-holes, narrow inlets, arches, gullies, buttresses and stacks are created. Some of the stacks offshore are formed into pinnacles populated by sea birds. Where the old red sandstone occurs on our coastlines, such as at Caithness, Pembrokeshire and elsewhere, this forms some of our most colourful and beautiful cliff scenery.

VARIOUS CLIFF TYPES

In addition to the foregoing basic cliff types, the British coastline also has other types, sometimes only occurring for a short distance locally, such as millstone grit, shale, slate. The coastline also frequently comprises a combination of several types of rock and other material, such as limestone and shale, sandstone and shale, chalk and clay, etc. A clue to these is the fact that the soft rocks will be eroded more rapidly than the hard rocks, thus the latter will still be forming jutting headlands while the soft rocks have receded to form inlets, coves and bays. The sea's removal of the soft rock or material, where both hard and soft rock or material are sited in bands in close proximity with each other, particularly if the soft rock or material forms the base of the cliff or shore, also means the collapse or destruction in a landslide of the hard rock, too.

COAST AND SHORE: BARS, SPITS AND DELTAS

A bar is a bank of sand, gravel or other material that has accumulated where the outward flow of a river meets the incoming tide, causing both to be checked, so that they both deposit their content of silt, sand, etc. This deposit gradually builds up until it either diverts the direction of the river's flow or forms a ridge or bank across the river mouth, as a harbour bar, at a distance offshore. Sometimes at a further distance fom the shore, in shallow or shoal water, the tide may create a similar ridge or bank known as an offshore bar or shoal. Surf waves breaking along the bar or shoal, the latter word being a variant of shallow, will be a clue to its dangerous presence. Both can be extremely hazardous to vessels, particularly during bad weather, and dredgers have continually to maintain a free passage. If the sand or silt is loose and heavily saturated with water it creates the treacherous quicksand that engulfs anyone who attempts to cross it. The building up of a bar sometimes also dams the outlet of a stream so that a marsh, lagoon or pool is eventually formed behind it.

Alternatively, a spit is a long, narrow area of deposited material, bank of sand, shingle, etc., which extends from the land into the sea, unlike the bar, which is parallel to the land. Sometimes sand and pebbles collect through the agency of the sea to such a degree on a spit that conditions are suitable for Marram Grass to become established, stabilising the spit, which eventually becomes an island or a promontory attached to the mainland. Following this many other species of flora and fauna are able to colonise the site.

A delta is a low-lying triangular tract of soil at the mouth of a river which enters the sea through two or more channels. This accumulation of soil is formed of silt carried along by the flow of the river until the point where the river is restrained by the incoming tide. The silt is deposited there and continually added to. Most British rivers are not large enough, or have too fast a flow, to form the huge deltas that exist in other parts of the world; also the regular tidal flows take away or reduce the amount of silt deposited. But one example is the wide flatland alongside the lower Thames.

ESTUARIES

An estuary is the broad mouth of a large river where it widens and meets the tides before reaching the sea. A famous example is the Thames Estuary. The physical structure of an estuary, however, can vary with each example, depending on the flow of fresh water into it, the silt this is carrying, and the degree of salinity, among other factors. Usually the water is completely fresh near its head and completely saline at its mouth, the salinity varying in-between the two. In some instances the estuary is almost the sea itself, especially at high tide. In the estuary or bordering it there may be mudflats, islands of vegetation, submerged at high tide, small creeks and inlets and salt marshes. The lowered salinity of the water where fresh and salt water meet and mingle determines the type of fauna found in estuaries. There is marine life which is able to adapt to conditions of varying salinity, or marine life which can survive only in brackish water areas of low salinity, and there are fish species which during part of their life cycle live in the sea and return to the rivers or vice versa, passing through the estuary while doing so.

MUDFLATS

A mudflat is an area of broad, level foreshore, in estuaries, formed of mud of varying liquidity. Where sand and gravel are mixed with the fine soil particles it is fairly firm, but if these are absent it is either a soft or very slimy, clogging ooze into which the feet and, in some places, the limbs sink. Interspersed in the mudflat area there may be small islands of vegetation, sometimes submerged at high tide. Such flats very often have numerous channels where water drains from the nearby land or are cut by the incoming tide, where fish lie at low water. The exposed mudflats are also a gathering place for large numbers of sea birds, Gulls, Waders, etc. If the mud is interspersed with sand, or is soft yet bears only partially decomposed material, various specialised marine life can exist, sometimes abundantly, but in the layers where decomposition is complete, so that the mud is black and stinking, no life is supported therein.

ROCKY SHORES

These vary according to the type of soft or hard rock from which they are derived and may, but not always, have come from a nearby cliff from which the sea and weather eroded them, to crash on to the shore and give some short-term protection to the cliff base. These shores may be littered with huge boulders or be carpeted with numerous, various-sized smaller portions. The amount of marine life present also varies, depending on whether the rock is rough – or smooth – allowing animals or plants to attach themselves, whether the shore is horizontal or sloping, and whether the rock has deep or shallow pools. There is no doubt the rocky shore has the greater lure for the naturalist seeking the often-beautiful fauna of the pools, and for the ordinary person, perhaps with children, who turns over rocks and stones just to see what is to be revealed underneath.

SALT MARSHES

A salt-marsh or salting, as it is also known, is a meadow or pasture close to the shore, rich in salt, and thus a good grazing for cattle. It may be land which in the past was won back from the sea. In some cases this was begun by the driving in of posts from the land to a distance offshore, sometimes only to act as a breakwater, boundary line or to denote a place to come ashore after fishing. These posts had the effect of forming a barrier to the tide, which is slow moving over a sloping shore anyway, and so the water dropped the silt it was carrying at this site. Dense layers of seaweed fringing a shore have the same effect.

This silt in time builds up to become heaps of mud, which, in turn, form a barrier so that more silt is deposited and so on. The heaps of mud become islands with channels through which the water wends its way. On the top and sides of the islands grow seaweeds, which are habitats for marine creatures, the latter attracting sea birds. The seaweed creates a further barrier so that more silt is deposited at the top and sides, which either enlarges them singly or links the islands to form a larger expanse. Eventually this process raises the islands above high-water level and the upper seaweeds disappear. The birds drop their excreta which, with the mud, provides a habitable situation for the plants whose seeds are carried to the islands on the feet of birds borne there by the tide or wind. As these grow and spread they stabilise the mud and themselves seed or vegetatively extend to continue their species. Man may now take over the sea's work by creating dykes and draining the marshes, perhaps building an earth or stone seawall to bar the sea, so that what was once a sloping area of shore is now valuable grazing or arable land. The draining also causes a further change, the salt-loving flora colonists now mostly disappearing, perhaps surviving in the dykes, ditches and brackish water pools, and being replaced with inland "weed" species or more useful grasses. Examples of this are to be found in the Medway Estuary. It is also possible to find various stages of this process, islands of mud being created, the first plants taking root on newly-risen islands, still continuing in suitable areas.

SAND DUNES AND SHORE

Sand comprises the loose, minute grains and particles of minerals created by the disintegration of rocks. It usually consists of quartz, which is much harder and more durable than the other minerals in rocks, the latter being continually reduced to even finer, smaller particles and carried away to form mud. Sand varies in colour depending on the dominant minerals but on the familiar seashore it is either a red-yellow, whitish-yellow or pale brown. The colour also changes according to its water content, sometimes being almost white when dry. Because of the water retention between the grains a large number of marine animals can live partially or completely within the sand, on the middle and lower shore. Sand also occurs in estuaries and river beds, where it has been carried by the water, as well as along the sea coast.

A sand dune is an area of sand, usually near the seashore, but sometimes further inland, and may be in the form of continuous ridges or mounds. The shape also depends on the wind. If it varies then the dunes will be shapeless mounds and ridges, but if the wind is in a constant direction they may have a definite crescent shape. Dunes are created by the wind removing and carrying the light, dry grains of sand from exposed sand flats until the wind falls or meets an obstacle and deposits the sand grains. When this occurs regularly dunes are created. In past centuries much valuable land, churches, buildings and other property have been engulfed on some parts of the coastline, so that the area resembles a miniature desert. Today the creation and extension of dunes can be restricted by the planting of Marram Grass, which can thrive in this seemingly barren habitat. The Marram spreads laterally while its roots penetrate deeply, sometimes as much as 50 feet. The spreading grass holds the sand surface together against the force of the wind's action, while the network of roots stabilises the depth of the dune. This stabilisation allows other plants which follow, such as Sea Spurge, Sea Holly, Sea Convolvulus, Wall Pepper, Seas Sandwort and various Grasses to become established. When this stage is reached, unless fresh sand is regularly added to the dune the Marram Grass loses its vigour and remains only in patches or disappears. The new dunes are whitish in colour, but as they age, with the addition of humus, animal and bird droppings, etc., they change to a brownish or greyish. Sand dunes can vary considerably in height, some examples of the older dunes being up to 50 feet.

SHINGLE AND PEBBLE SHORES

Shingle is a mixture of varying-sized stones and pebbles. Pebbles are smallish stones rounded or semi-rounded. These were originally portions of rock from either erosion of the cliff face or sea bed, which through centuries of wave action and rushing, sometimes violent, movement of the sea have become reduced and fragmented in shape and size. The pebbles similarly have been ground together by the sea's action and rounded. Among the pebbles there may be other stone material irregular in shape and in the process of being ground into round, smooth shapes or breaking down into much smaller particles. The pebbles and stones may be unlike the local cliff or coastline because they have been brought to the site from somewhere else on the coast, either by drifting along the shore or being carried by tidal currents. Usually the shingle shore has a steep incline up which shingle, gravel and sand are carried by the incoming dash of the waves to drop back as the water recedes. This continual grinding movement of the material ensures that no marine life can exist on or in this part of the shingle shore, the only evidence that marine life does exist in deeper water here being the occasional piece of seaweed or luckless crab flung up the shingle. On the level area of the shingle shore, above high water mark, however, which may have been created years before and added to, where conditions have become stable through the sea receding, a characteristic flora and fauna can establish itself. Some of the more notable shingle areas are at Dungeness, Kent, Chesil Beach, Dorset, Orford Ness, Suffolk, Slapton Sands, Devon, The Crumbles, Sussex.

Chapter 2

Birds

The best known birds of the coast and estuary are the "seagulls", probably because they are the most frequently seen. There are, however, a large number of other birds which occur in these habitats. Certain species do so all the time, others only during the winter or summer, when on migration or when nesting and rearing their young.

In addition to these which are found always or usually close to the sea, the visitor will soon notice many inland birds that also go down to the shore to feed if within range of it. Starlings very often do so, probing and searching for any titbit along the upper shore of estuaries, mudflats and coasts or on salt meadows. Rooks do the same. House Sparrows may frequently be seen on the seashore if their usual home is a coastal village or resort, or fields and meadows close to the sea. Herons regularly return to the seashore or to brackish water and salt marshes near the shore. Swans, too, will occur on tidal rivers and estuaries.

In hard winters species that normally frequent inland waters and rivers go to the coast for food. Kingfishers sometimes can survive only by doing so, feeding on the marine life they find there. As I refer to in Chapter 18 many migrating inland birds stay a short time to rest on the coast when arriving here or departing to or from Scandinavia, the European continent or to Africa. But this chapter must, for space reasons, deal with only the most frequently seen or particularly interesting species of coast, estuary and sea birds.

To aid the beginner in identification I have used various terms which define the type of each bird, so that it can be ascertained when it is most likely to be seen. As most bird-watchers know there are exceptions to the rule and birds, being mobile and having wills of their own, do arrive unexpectedly in the unlikeliest of places to contradict all the text-books. But normally the species fall within the following classifications—

"Resident" is a bird which resides in the habitat for all or most of the year and nests there;

"Summer visitor" is a bird arriving here from overseas to nest, from early spring onwards to late summer, after which it vacates the habitat in autumn;

"Winter visitor" similarly is a bird arriving from perhaps the Arctic regions, Russia or Scandinavia, to spend our milder winter months with us from late September until Spring;

"Passage migrant" is a bird that passes through the British Isles during its migration, either in spring or autumn, when it rests and feeds and can be seen before continuing its journey. It will be noted that some are "resident" and "winter visitor" or "summer visitor", which indicates that the British population of that species is supplemented by others arriving for that period of time from the Continent, Scandinavia, etc. Also "passage migrant" and "winter visitor" is a bird that usually migrates along the coast, but afterwards in winter other examples of the species also arrive and stay or continue their journey.

THE WADERS

This group of birds, with several exceptions, occurs in shallow water or at the water's edge in estuaries, on coastal seashores, muddy, sandy flats and mudflats, where they actively wade and run about, using their bill to probe in search of food. Their diet varies, but includes insects, worms, crustaceans, molluscs, fish, seashore vegetation. When the breeding season is over some of them merge into huge flocks in autumn and winter, particularly if migrating, sometimes several species intermingling as they, somewhat restlessly, fly, wade or swim about the habitat. While doing so some of them have conspicuous features which aid identification. To identify other species various differences must be noted, such as coloration of plumage, wing-patterns, body size, shape of bill, or some characteristic habit of wading movement or flight.

BAR-TAILED GODWIT (*Limosa lapponica*)

Passage migrant in spring, autumn, and a winter visitor. It frequents estuaries, mudflats and sandy coastal flats. It is a mottled browny-grey on the upper-parts, the tail being barred brown-grey and white, the rump and underparts being dull white. The long legs are grey, the long bill being slightly, but obviously, curved upwards. When in flight the legs do not extend beyond the tail and the white rump appears as a conspicuous patch above the tail. It measures about 16 inches in length, including the 4 inch bill.

BLACK-TAILED GODWIT (*Limosa limosa*)
Passage migrant in autumn. It frequents estuaries, mudflats, also sewage farms, salt marshes and inland lakes. Although formerly rare it has increased in numbers in recent years and regularly spends the winter in Britain. It is darker than the Bar-tailed Godwit, being a dark mottled brownish-grey, with a broad white bar on the wings that is prominent when in flight. The tail is white with a wide, black band at the tip, but the rump is dark grey and the underparts light grey. The greyish-black legs are also longer than the last described and in flight the feet do project beyond the tail. So, too, is the longer bill which is usually straight or only slightly curved at the tip. It measures about 17-18 inches, including the bill.

COMMON CURLEW (*Numenius arquata*) Fig. 2
Resident, also migrant from the Continent. It commonly frequents mud-flats, estuaries, salt marshes, sandy flats and rocky coasts. The plumage

Fig. 2 Curlew *Numenius arquata*

is brown with small dark brown or black streaky marks, with whitish underparts and the rump has a white patch which can be seen when the bird is in flight, but the best aid to identification is its long legs and very long bill, about 5 inches, which is down-curved. It often wades at a considerable depth into the water and is a very shy, wary bird. The familiar, plaintiff call "cour-lee-cour-lee" helps to determine its presence and identify it from its relative, the Whimbrel. It is our largest wader, being 23-25 inches in length, including its bill.

Fig. 3 Redshank *Tringa totanus*

COMMON REDSHANK (*Tringa totanus*) Fig. 3
Resident, also arrives from the Continent in autumn. It very commonly frequents, sometimes in large flocks, mudflats, estuaries, salt marshes, sewage farms, reservoirs, sandy shores and dunes. It has greyish and brown upper-parts, a streaked throat and upper breast, white under-parts, the rump being white, with black and white on the tail and a broad white mark on the rear of each wing, the latter being very prominent when the bird is flying. The legs, from which it derives its name, are red-orange, while the straight black-tipped bill is red-orange at its base. In some conditions this bird may appear to be black and white. It measures 11 inches in length.

COMMON SANDPIPER (*Tringa hypoleucos*)
Summer visitor and passage migrant. It frequents estuaries. The upper-parts are brownish-grey or olive-brown, the breast is ashy-grey, under-parts white, with a white bar on the wings which is also prominent in flight. Its bill is dark brown, the legs being greenish. It is an active bird, running on the shore and when it halts has the habit of dipping its head and jerking its tail in an up and down motion. When it flies, with a rapid beating and gliding movement, its wings appear strangely bent as it swiftly passes low over the water. It measures about 7½ inches in length.

COMMON SNIPE (*Capella gallinago*)
Resident. It commonly frequents salt-marshes and freshwater marshes, pools and meadows within proximity of the shore, also sewage farms and mudflats. Usually singly in summer, but sometimes in flocks during winter. It has a rich mottled black and brown plumage, with a longitudinal barred appearance, white underparts and central lengthways light stripes on the crown. The plumage contrasts so well with its background it can almost be stepped on before rising into a zig-zag flight and making a "hoarse" alarm note at the same time. The straight bill is long, about 2½ inches. From early spring until summer the male "drums" or "bleats". This he does by rising in circles to a considerable height, then diving towards the habitat at speed with the tail feathers stiffly spread out like a fan. The wind against and passing the outer feathers vibrates them and creates the "drumming" or "bleating" sound. It measures about 11 inches in length, including the bill.

DUNLIN (*Calidris alpina*) Fig. 4
Resident. It frequents estuaries, mudflats, salt marshes, sand banks and seashores. Probably it is the commonest small wader. Sometimes it occurs in huge flocks which perform acrobatic manouvres when in flight, or accompany other waders. In winter the upper-parts are streaked brownish and grey, with a white wing bar, the throat and underparts white with a greyish breast band, but the plumage is very variable. The bill is black and straight or slightly decurved while the legs are dark olive. In summer it has a large black patch over the lower breast, a speckled throat, and the upper-parts dark to chestnut-brown and black. It measures about 7 inches in length.

KNOT (*Calidris canutus*) Fig. 5
Passage migrant in autumn and winter visitor. It commonly frequents

Fig. 4 Dunlin *Calidris alpina*

mudflats, estuaries and sandy flats and coastline. It often arrives in huge flocks in these habitats where the birds remain together, all heads facing in the same direction, "working" forwards in unison when probing for molluscs and worms, staying close if alarmed into flight, when the flock moves almost like a dense grey cloud. In winter plumage the upper-parts are light grey, underparts being white with grey-brown streaks; the tail has black bars and part of the rump is white, while the straight bill is black and the legs green. It measures 10 inches in length.

Fig. 5 Knot *Calidris canutus*

LAPWING, GREEN PLOVER OR PEEWIT (*Vanellus vanellus*)
Resident and passage migrant, summer and winter visitor. Large numbers arrive from the Continent to supplement our resident birds. It frequents mudflats, in estuaries, on salt marshes and meadows, also cliffs. In winter flocks come inland to arable land and meadows to feed and in many cases stay inland throughout the summer also. The upper-parts are dark metallic green, the throat and breast are black, the rump and neck sides white and underparts white. The bill is short and black, the legs brown. An aid to identification is the long, greenish-black crest on the head. In flight the plumage appears black and white; also characteristic is the way the broad, rounded, blunt wings beat in the slow tumbling flight. When inland the Lapwings frequently all face the same direction in the field. They also may so blend against their background that they cannot be seen until they move, though their presence may be revealed by their somewhat sad, plaintive "pee-wee" calls. It measures 12 inches in length.

OYSTER-CATCHER (*Haematopus ostralegus*)
Resident. It frequents sandy, shingly, pebbly shores, also mudflats, salt marshes, estuaries, rocky coasts and cliffs. It is very easy to identify. The upper-parts are a glossy black except for the white rump and bar on each wing, there being a black band on the edge of the white tail, the underparts are also white. It has a long, straight orange-red bill, the long legs and feet being flesh pink. It can open oysters, but despite its name, if it relied entirely on finding these molluscs for its food it would go hungry, and so preys on mussels, limpets and other molluscs, as well as crustaceans, worms and insects. It measures 17 inches in length.

Fig. 6 Ringed Plover *Charadrius hiaticula*

RINGED PLOVER (*Charadrius hiaticula*) Fig. 6
Resident, also passage migrant. It commonly frequents estuaries, mud-flats, sandy dunes and shores, salt marshes, shingle and rocky coasts, sometimes in large flocks. The upper-parts are greyish-brown, the back and crown of the head being dark brown, the underparts white. Other areas of the head are black and white, with a white area on the "fore-head", there also being a wide dark band through the eyes. An aid to identification is the broad, black band across the white breast like a gorget, below its white "collar". On the wings there is a white bar. The short bill is orange-yellow with a black tip, legs being yellow. It measures about 7½ inches in length.

TURNSTONE (*Arenaria interpres*)
Passage migrant and winter visitor. It frequents shingle, stony and rocky coasts, sandy, pebbly flats, occasionally mudflats. In summer the upper-parts are red-brown and black giving the bird a "tortoiseshell" appearance, with a broad, very dark mark across the breast, the rump being white and underparts entirely white. On the head and neck it has clear black and white markings. In winter the upper parts are brown and it loses its "tortoiseshell-like" plumage. The short, black, slightly upturned bill is pointed at its apex and the legs are orange. In flight this bird has a distinct pied appearance and yet almost disappears from sight when it alights on a rocky, stony shore. An aid to identification is its characteristic feeding habit of continually turning over stones, pebbles, shells, debris and seaweed with its bill in search of marine life, hence its name, although it also opens mussels and limpets. It also digs in the sand and shingle for sandhoppers. It is about 9 inches in length.

WHIMBREL (*Numenius phaeopus*)
Passage migrant. It frequents mudflats, estuaries, salt marshes, sandy flats, rocky coasts and shores. Its streaked brown plumage is similar in appearance to the Common Curlew, but is smaller, the bill is shorter and less curved, also it has broad, dark brown bands divided by a pale stripe in-between over the crown of the head. It measures about 16 inches in length, including the 3-inch bill.

THE DUCKS

The basic appearance of our wild ducks is similar to the colourful domestic ducks which are descended from a wild duck, the Mallard. They fly, like geese and swans, with the neck and head outstretched, their smaller, narrower wings beating rapidly, but able to achieve skilful movement and it is sheer beauty to see numerous ducks leaving or alighting on their habitat. Ducks naturally spend more time on the water than the larger geese and differ in the way they do this. The Diving Ducks, as this classification indicates, regularly dive in the water to search for food. They are usually larger birds and walk erect and ungainly because their legs are positioned well back on the shorter body. The Surface Feeding Ducks sit higher on the water and obtain food by "up-ending" with their head and breast under the surface and tail pointing skywards, or they "dabble" by groping on and below the surface with their bill. If on water ducks should be watched to note their method of feeding which helps to classify the species for easier identification. They feed on fish, crabs, and other marine life, some vegetation, also feeding as they walk about on the land habitat.

The males are usually much more brilliantly plumaged than the female, the latter usually being dull in appearance and the females of some species resemble each other, so that small differences have to be noted. Identification is further complicated during the summer when the male ducks assume what is known as their "eclipse" plumage so that during moulting they may resemble the females, except for the retention of markings on the wings. Classified with the Diving Ducks, called the "Sawbills", is the Goosander, Red-breasted Merganser and Smew, which dive for fish, because the edges of the bill mandibles have saw-like "teeth" pointed backwards to capture and hold fish. The Shelduck, although usually described with the ducks, is more closely allied to the geese by its appearance, with both sexes, unlike other ducks, being similar, and in their walk and flight.

COMMON EIDER (*Somateria mollissima*) Fig. 7

Resident and winter visitor. From May it frequents rocky coasts and

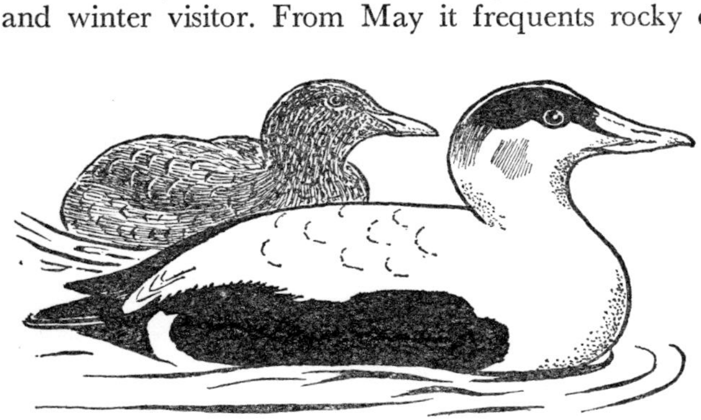

Fig. 7 Eider Duck *Somateria mollissima*

islands, sometimes in colonies, and it is more maritime than other duck species. A diving duck. The male upper-parts, head and neck are white, the crown, tail and underparts being black, the nape of the head being greenish and a rosy-pink tint on the white breast. It also has prominent white fore-wings. The female is entirely red-brown with numerous, narrow, black bars and mottling. An aid to identification is the unducklike bill, the upper mandible surface of which is straight forming a line with the forehead and not curved or knobbed, as with other ducks. This bird uses a large amount of down from its breast to form a soft lining for the nest and from this arose the term "eider-down" for the bed-quilts formerly stuffed with this material. It is our largest British duck, measuring about 24 inches in length.

COMMON SCOTER (*Melanitta nigra*)
Resident and winter visitor. It frequents coasts, particularly offshore, sometimes in large numbers, as it is a maritime duck. A diving duck. The male is completely black, except for the bill which has an orange patch, and also has a swollen black "knob" at its base. The female has dark brownish upper-parts with a lighter, greyish area on the side of the neck, brownish-white underparts while the bill is more greenish-black. This species has pointed tail feathers and while swimming elevates these. It measures about 20 inches in length.

GOLDEN-EYE (*Bucephala clangula*)
Winter visitor and passage migrant. It frequents estuaries and coasts, also coming inland to lakes and large stretches of water. A diving duck. The male upper-parts, tail and triangular-shaped head are black, breast and underparts white, there being a large, roundish, white patch below the eye near the black bill, the aid to identification. On the wings there is a square white patch which is prominent when the bird is in flight and swimming. The female has mottled greyish-black upper-parts with a chocolate-brown triangular-shaped head, a white collar and wing patches. Both male and female have golden-yellow eyes, hence their name. It measures about 18 inches in length.

GOOSANDER (*Mergus merganser*)
Resident and winter visitor. It frequents estuaries, mudflats and inland water, reservoirs, etc. A "Sawbill" diving duck. The male has a black

back, glossy dark green head, white breast and underparts with a pinkish tinge, and white wings. The legs, feet and narrow bill are bright red. The female has bluish-grey upper-parts, dark red-brown head and neck and white breast and underparts with a prominent white patch on the throat, also pinkish-red legs. Both sexes have a crest on the back of the head but it is difficult to see, especially when this is lying prostrate. It measures 26 inches in length.

LONG-TAILED DUC K(*Clangula hyemalis*)
Winter visitor. It frequents offshore water usually. A diving duck. The male has a black or very dark brown back, breast and wings, underparts white, the head and neck being white with a black or dark brown spot mark on the side of the neck. The pink and black bill is short and the head looks rather small for the size of the body, accentuated by the two very long central tail feathers projecting five inches or so beyond the rest of the tail. These he either carries erect or trails them. The female has dark brown upper-parts, white underparts except for a brown breast mark, and a white head with a brown crown and mark on the neck, her tail being without the long feathers. The male is about 20 inches in length including the tail feathers, the female is about 17 inches in length.

MALLARD (*Anas platyrhynchos*)
Resident. It frequents estuaries, bays, mudflats, salt marshes, sea coasts, as well as inland habitats. A surface feeding duck and probably our most common duck species. The male has a glossy metallic-green head and neck, with a white ring below this around the neck like a collar, the upper-parts are greyish-brown, the breast chestnut-brown with brownish and grey underparts and a yellow bill and orange legs. On each wing there is a purple-violet patch and two white bar markings. It is such a colourful and handsome bird it is almost impossible to mistake it. The female is much duller, being brown with darker markings and orange-brown bill and orange legs, the patch and bars on her wings being conspicuous. It measures 22 inches in length.

PINTAIL (*Anas acuta*)
Resident and winter visitor. It frequents estuaries, mudflats and sea coasts. A surface feeding duck. The male is greyish from a distance, but has a grey back and flanks, white underparts, and black under the tail with a white or yellowish spot in front of it. The head and neck are chocolate-brown with a prominent white stripe down the side of the neck joining the white breast. The aid to identification is the long, elevated, pointed tail, from which it gets its name, which is also notice-

able in flight. On the wings it has a bronze-green patch. The female is brownish and not unlike the female Mallard, but the female Pintail is more slender, with a grey bill, and has a pointed, but not elongated tail. In length it measures about 22 inches.

RED-BREASTED MERGANSER (*Mergus serrator*)
Resident and winter visitor. It frequents estuaries, mudflats and bays and offshore waters being almost a maritime duck. A "Sawbill" diving duck. The male is a handsome bird having a glossy blackish-green head and upper neck, a white "collar", black back, a wide patch of chestnut on the breast, white underparts and grey flanks. The shoulder has prominent white and black bars and white marking on each wing. The bill and legs are red. On its head it has an erect divided crest. The female has brownish-grey upper-parts not unlike the female goosander, but also has the prominent erect crest. It measures about 23 inches in length.

Fig. 8 Shelduck *Tadorna tadorna*

SHELDUCK (*Tadorna tadorna*) Fig. 8
Resident. It frequents estuaries, mudflats, dunes, sandy and other coasts, salt marshes. A surface feeding duck. This species is more goose-like than duck-like in general shape, movement and habit, such as its slow flight. The male has a dark green head and upper neck, upper breast white with a chestnut-brown band below around it, the upper-parts being white except for black and brown on the wings, the underparts

also being white divided by a central blackish stripe. The bill is bright red with a prominent "knob" at the base during the breeding season. The legs and feet are pink. The female is very similar to the male except she is smaller and not quite so brilliantly plumaged and the bill "knob" is absent. It measures about 24 inches in length.

SMEW (*Mergus albellus*)
Winter visitor. It frequents coasts and estuaries, sometimes salt marshes and inland water. A "Sawbill" diving duck. The male is entirely white except for a black mark under each eye near the bill, pale grey flanks, and a black band down the back. This coloration is an aid to identification as he is the whitest wild duck. The female has grey upper-parts and breast, white underparts and mark on the wings, a red-brown crown and white cheeks and throat. The male and female have a small crest on the head. It measures about 16 inches in length.

TEAL (*Anas crecca*) Fig. 9
Resident and winter visitor. It frequents estuaries, salt marshes and the sea coast, sometimes in large numbers. A surface feeding duck. The male has upper-parts and underparts greyish with brown markings, but the head colour pattern, which looks rather oriental, is an aid to identification. The crown, back of the head and side of the neck and cheek are dark chestnut-brown, but around the eye and behind it down the nape is a curving patch of bright glossy green which is enclosed within a narrow light brown border. On each wing there is a purplish-black and green patch and a long whitish streak. The female is not unlike the female Mallard, being brown with light brown markings and similar wing markings as the male. This species is the smallest British wild duck, measuring about 14 inches in length.

Fig. 9 Teal *Anas crecca*

WIGEON (*Anas penelope*)
Resident and winter visitor. It commonly frequents estuaries, mudflats, salt marshes and the coast, also inland stretches of water. A surface feeding duck. In some places it may outnumber the Mallard as the commonest British duck in the area. The male has greyish upper-parts, with brown bars, a pinkish breast, chestnut-red head and neck, a creamy-buff crown and forehead. On the flanks it has a prominent white patch and a black and green patch and white patch on the wings. The female is duller, greyish-brown on the upper-parts and lighter buffish-brown on the underparts, with a white patch on the wings. It measures $18\frac{1}{2}$ inches in length.

THE GEESE
Wild Geese have a similarity to farmyard geese on the ground. Several species occur on the habitats in Britain during the winter. They are usually seen in their V-formation or irregular flight pattern and making their "honking" or other conversational vocal notes, depending on the species. It is a thrilling sight and sound.

Fig. 10 Barnacle Goose *Branta leucopsis*

BARNACLE GOOSE (*Branta leucopsis*) Fig. 10
Winter visitor and passage migrant. It frequents estuaries, mudflats, salt marshes, turfy coasts and near the sea. The upper-parts are bluish-grey and black, with the neck and breast black, the face and forehead

white, the underparts being greyish-white, with a white rump and black tail. The legs are black, also the short bill is black. The black and white coloration of this goose is an aid to identification. It measures about 23-25 inches in length.

BRENT GOOSE (*Branta bernicla*)
Winter visitor. It frequents estuaries, mudflats and rocky coasts, sometimes in large numbers. The upper-parts are dark grey, the head, neck and breast are black with some small white patches on the neck that may form a "collar", the underparts being dark grey and brownish, with white tail coverts. There are two types: one has a dark breast, the other light coloured. It is the smallest wild British goose measuring about 22–24 inches in length.

GREY LAG GOOSE (*Anser anser*)
Winter visitor and passage migrant, chiefly to northern Britain. It frequents estuaries, mudflats, salt marshes, turfy places near the sea, also inland on moors, marshes, boggy land and loch islands. The upper-parts and underparts are grey-brown, including the head, except for the rump, tip of the tail and underside of the tail which are white. The front edge of the wings is bluish-grey, the legs and feet being flesh pink. The bill is yellowish-orange with the tip of the upper mandible white. It is our largest wild goose measuring about 34–36 inches in length.

PINK-FOOTED GOOSE (*Anser brachyrhynchus*)
Winter visitor and passage migrant. It frequents the coast, estuaries salt marshes and grassy places near the sea, sometimes occurring in large numbers. It has bluish-greyish upper-parts, dark brown neck and head, brownish breast and white on the tail underside. The legs and feet are flesh-pink, hence its name. The short bill is black except for the middle area of this which is pink, the tip of the upper mandible also being black. It measures about 28 inches in length.

WHITE-FRONTED GOOSE (*Anser albifrons*)
Winter visitor. It frequents estuaries, salt marshes, bogs, coasts, and grassy places near the sea. The upper-parts are dark brownish-grey, the breast having some irregular black bar markings, but the aid to identification is the prominent white patch on the forehead and front of the face above the bill which is pinkish with a white tip to the upper mandible. The legs and feet are orange. It measures about 27–30 inches in length.

1 An above view of the Compass Jellyfish *Chrysaora hyoscella* stranded on a sandy beach.

2 Coltsfoot *Tussilago farfara*, an early Springtime flower, which colonises sandy habitats close to the shore when the latter has been stabilised by the root systems of various Grasses.

3 A colony of Acorn Barnacles *Balanus balanoides*.

4 A side view of a Compass Jellyfish *Chrysaora hyoscella* swimming. This species has stinging cells that penetrate human skin and so should not be touched.

5 A Common Dog Whelk *Nucella lapillus* with a banded shell, one of several variations.

The aerial grace of a Kittiwake *Rissa tridactyla* leaving its colony on a steep cliff face.

An adult Herring Gull *Larus argentatus* with young.

THE GULLS

The general appearance of the Gulls is well-known. A little study of the various species and their differences in size, coloration, habits or habitat should make it a fairly simple matter to identify each of them if met with. They are mostly excellent "nature's dustmen", scavenging far out at sea on garbage, feeding on refuse on the shore, around harbours, ports, etc. In addition they eat dead or alive marine creatures, worms, insects, also seaweed, some taking other birds' eggs and young during the breeding season. They inhabit coasts, estuaries, mudflats, salt marshes, as well as the seashore. Some species are a familiar sight inland, following the plough, on rivers, even in towns and cities near water where they are almost classed now as land birds. Certain of the British resident gulls' numbers are supplemented by migratory invasions of their species from Europe and further afield.

BLACK-HEADED GULL (*Larus ridibundus*) Fig. 11 (a)
Resident. Very common, particularly on the coast, in estuaries, on mudflats, also inland following the plough or in grassy fields and in and around habitations near the sea, sometimes in large, noisy flocks on sewage farms and around sewer outlet pipes. During the summer the adults have a chocolate-brown, *not* black head, crimson-red bill, feet and legs. An aid to identification is the black tip to the grey, pointed wings, which also have a conspicuously clear white margin on the front edge. In winter the head of the adults is white except for a dark brown or blackish mark or marks before, behind or near each eye. It measures 15 inches in length.

COMMON GULL (*Larus canus*) Fig. 11 (b)
Resident. Far from common, occurring mainly on the coast, sometimes in estuaries, either singly or in flocks. It can only be said to be common in Scotland and parts of Ireland. The adult has a white head, pale grey back and wings which have black tips with white marks and spots. The bill, feet and legs are yellowish-green. It can be distinguished from the Herring Gull, with which it may be confused, by the Herring Gull having flesh-coloured legs and feet, a yellow bill with a red spot and is bigger. The Common Gull adult has brownish-grey streaks markings on its head in winter. It measures about 16 inches in length.

GLAUCOUS GULL (*Larus hyperboreus*) Fig. 11 (c)
Winter visitor. It occurs chiefly in northern and eastern England and Scotland, though may come as far south as East Anglia, frequenting low-lying coasts, sea cliffs, estuaries, harbours, etc. The adult has a white

Fig. 11 The Gulls
(a) Black-headed Gull *Larus ridibundus*
(b) Common Gull *Larus canus*
(c) Glaucous Gull *Larus hyperboreus*
(d) Great Black-backed Gull *Larus marinus*
(e) Herring Gull *Larus argentatus*
(f) Kittiwake Gull *Rissa tridactyla*
(g) Lesser Black-backed Gull *Larus fuscus graellsii*

ead, yellow bill with a red spot, a lemon-yellow ring around each eye, nd flesh-pink legs. The back and wings are light ash-grey, the tips of ıe wings, tail and underparts being white. There is *no* black at all on ıe wings which is an aid to identification from the smaller Herring ull. In winter the adult Glaucous Gull has pale brown streaks on its ead and underparts. It is about 25–28 inches in length.

REAT BLACK-BACKED GULL (*Larus marinus*) Fig. 11 (d)
esident. Uncommon. Occurs more often at sea or offshore than over nd, except in winter, then on coasts and estuaries, and during the eeding season, when on rocky coasts, cliffs and islands; usually singly ıt sometimes in small flocks. The adult has a black back and wings, ith white marks near the black wing tips; head, tail and underparts hite, the legs and feet being a pale flesh colour. The massive, powerful ll is yellow with a red spot. It is a winged pirate, its size and strong, illed flight allowing it to pursue and rob other birds of their prey and will raid their nests for eggs and young; also it kills numerous sea rds as part of its diet. In winter it has a few brown streaks on the head d neck. It is the largest British gull, between 27 and 30 inches in ıgth.

ERRING GULL (*Larus argentatus*) Fig. 11 (e)
sident. Common, particularly frequenting around the coasts, on the ore or cliffs, also in estuaries, on mudflats, salt marshes, usually in isy flocks, also inland. The adult has a pale soft grey back and wings, wings having black tips with white spots, and the head, tail and derparts are white. The legs and feet are flesh pink while the strong is yellow with a red spot. In winter the white head and back of the k have brown streaks. Despite its name it feeds on marine creatures er than herrings, is a first-rate robber and has the habits of stamping prodding on the muddy sand to cause the worms to come to the face, and opens mussels and other molluscs by picking them up in beak and dropping the mussel from a height on to a hard surface, ugh it also swallows some mussels whole, then disgorges the in-estible shells. It measures about 22 inches in length.

TTIWAKE GULL (*Rissa tridactyla*) Fig. 11 (f)
ident and winter visitor. It frequents precipitous rocky cliffs in nies in the breeding season, otherwise at a considerable distance from l, usually singly. The adult has a soft grey back and wings, with re black tips to the wings, the legs and feet being brownish-black. head, tail and underparts are white, the bill yellow. In winter the

head is a greyish colour. Graceful in swift flight and a good swimm it feeds either by diving for fish, crustaceans, vegetable matter, plankto when flying or swimming. It derives its name from its call notes "kitt wa-ak" and measures about 15–16 inches in length.

LESSER BLACK-BACKED GULL (*Larus fuscus graellsii*) Fig. 11 (
Summer visitor. Locally common. It frequents coasts, estuaries, mu flats, also occasionally inland, particularly during migration. Numbe of this species have stayed on during winter so it may be becoming resident. It more often occurs in larger flocks than the Great Blac backed Gull. The adult has a slate-grey back and wings, a white hea tail and underparts, the legs and feet being yellow and the bill yello with a red spot. In winter the head has black-brown streaks clo together while the back of the neck has blotch markings. It varies fro the Great Black-backed Gull in the colour of the legs, back and win and being smaller in size. It measures about 21 inches in length.

LITTLE GULL (*Larus minutus*)
Winter visitor and passage migrant. It frequents coasts, singly or small flocks. The adult has a black head, the back and wings are bluis grey, with white tips, the undersides of the wings being slaty-black, b the upper-side has no black coloration. The bill is brownish-red Summer, dark almost black in winter; the legs and feet are vermilic red. In winter the head is white, except for the crown, back of he and a blotch behind the eyes being grey. It measures about 11 inc in length and is the smallest British gull.

THE TERNS

At first glance the Terns may seem similar to some of the rela gulls, but on closer examination it will be realised that there are seve differences. During their breeding season, when present in summer the British Isles, they all have a black "crown" on their head, the upp parts are bluish-grey, and they have forked tails. From the latter ar their nickname of "Sea Swallows". They are also much more streamli than gulls, with very narrow wings and thin bills. Even if few of th details can be seen from a distance with the naked eye, their habit hovering over the water, like a hawk, then diving into it or swoop close to the water to sieze a fish or sand eel, or seemingly restle flying about when prey hunting, gives a clue that the bird or b belong to the terns. When not hunting they are graceful in flight, vhile doing so are difficult to identify for certain, because the chief lifferences between the terns are variations in their black crown, the :olour of their bill, legs and feet, and length of their tail feathers, so)inoculars are a necessity.

ARCTIC TERN (*Sterna paradisea*)
Summer visitor or autumn migrant. It frequents the coasts, offshore slands, sand dunes and beaches in colonies during the breeding season. The adult has a black crown and nape of the head, greyish back and ghter grey underparts, with a white rump and tail. The bill is blood-ed and the legs and feet crimson-red. So-named because it crosses the world from the Arctic region, where it breeds, as well as in the British sles, to the Antarctic Ocean, where at sea it passes our winter, which summer in the Antarctic.

COMMON TERN (*Sterna hirundo*) Fig. 12
Summer visitor or autumn migrant. It frequents the coasts, also estuaries, ıland lakes and reservoirs and tidal rivers, but breeds in colonies on lands, coasts, sand dunes and beaches. The adult is similar to the rctic Tern, but is paler, being less grey and more white on its under-arts, the outer tail feathers are also shorter and so do not extend eyond the wing-tips when closed. Its legs and feet are red. An aid to entification is the orange-red bill having a black tip in summer. It easures 14 inches in length.

Fig. 12 Common Tern (summer) *Sterna hirundo*

TTLE TERN (*Sterna albifrons*)
mmer visitor, or autumn migrant. It frequents coasts and sandy or ngle beaches and nests in colonies on beaches and dunes. The adult s a black crown with a broad white forehead and black eye stripe ıich are aids to identification. The bill is yellow with a black tip, the s and feet are also yellow. This is our smallest tern, being about 9 hes in length.

ROSEATE TERN (*Sterna dougallii*)
Summer visitor. It less frequently occurs around our coasts and is more maritime, but nests with other tern species. The adult has a black crown on its head, but the upper-parts are paler, whiter, than the Common and Arctic terns. In summer the bill is black with red at its base, which is an aid to identification. The breast has a rosy-pinkish colour that is sometimes difficult to see and from which it derives its name. The feet and legs are also red. If it can be seen perching it will be noticed the outer tail feathers stream longer than the tips of the closed wings. It measures 15 inches in length.

SANDWICH TERN (*Sterna sandvicensis*)
Summer visitor or autumn migrant. It frequents coasts but is chiefly maritime, excepting nesting in colonies on sand and shingle beaches offshore islands. The adult has a black crown with long, pointed, crest like feathers on the back of the head, an aid to identification. It i whiter than the other tern species, the slender, long bill is black with yellow tip, the legs and feet being black. The white tail is shorter an not so deeply forked, while the breast may be a salmon-pink hue. I was given its name because it was first found in the British Isles, a Sandwich, Kent, in 1784. It is the largest of the British terns, measurin 16 inches in length.

THE PETRELS AND SHEARWATERS

These two groups of birds are almost completely oceanic during the lives spent out at sea, except when they return to land for the breedin season or if blown inshore or inland by a storm. They deserve inclusio in this book, however, because some of them are well-known and the nesting around our coastline makes them eligible. At the base of the bills they have external tube-like nostrils.

FULMAR (*Fulmarus glacialis*)
Resident. It frequents cliffs and islands in colonies during the breedi season and has become increasingly common. The head, neck ar underparts are white, the back, narrow wings and tail being grey, wi darker grey on the wing tips. In appearance it could at first be mistak for a gull, but there are several differences, one aid to identificati being the wedge-shaped tail, another the yellow bill with tubed nostri Apart from these its sheer mastery of flight, holding the wings stiff a straight as it glides, then rapidly flaps its wings again and soars, swee

and ascends or descends on sometimes strong air currents high above or almost touching the waves, is an aid to identity. It measures 18 inches in length.

MANX SHEARWATER (*Puffinus puffinus*)
Resident. It frequents cliffs and coastal islands in colonies during the breeding season, but outside this period several individuals may be seen together. The upper-parts are black, the underparts white. It has long, straight, narrow wings on which it flies low above the waves in a very ast, gliding method of flight, occasionally rapidly beating the wings and frequently tilting them from one side to the other so that the watcher alternately sees white underparts then dark upper-parts. This is the nost common of the four shearwaters that occur around the British Isles, but, despite its name, it no longer occurs on the Isle of Man except as occasional migrant. It was given its name supposedly for the way it glides close over the sea, sometimes penetrating the surface with the wing tips. It measures 14 inches in length.

TORM PETREL (*Hydrobates pelagicus*)
Resident. It frequents rocky offshore islands in colonies in the breeding eason. It is very dark brown or black, except for the prominent white ump conspicuous during flight. The wings are long and when in flight ow over the waves it has a similarity to the house martin, but has the emarkable habit of "treading" momentarily on the water with its long gs and webbed feet as it flies close above the waves. Probably equally ell-known is its name of "Mother Carey's Chicken" applied by sea-arers who thought its appearance around their ship was a warning of ormy weather to come. It is not only our smallest web-footed bird but so the smallest sea bird, measuring about 6 inches in length.

HE AUKS
The Auk family are also almost completely maritime, but they do not nture so far from land as other species and remain in sight of it, turning to the coastal habitats for their breeding season, usually from ring to early summer. They are diving birds and pursue their prey, iefly fish, under water. Their flight is quick, sometimes skimming low er the sea on their comparitively small, short, rapidly-beating, slender ngs, but these are excellent for swimming. They walk erect, but addle ungainly due to their feet being positioned well back on their dy. Occasionally they are blown inland by strong winds, even reaching

counties far from the sea and unhappily today many hundreds of these birds become oil-coated and die horribly around our coasts.

BLACK GUILLEMOT (*Uria grylle*)
Resident. It frequents rocky coasts, cliffs and islands, sometimes on quite low habitats, chiefly Scotland and Ireland, but does not usually form huge flocks. Its summer plumage is black, with a large, prominent white patch on each wing. In winter it is white except for black mottling on the upper-parts and neck. The pointed bill is black, the legs and feet being red. It measures about 14 inches in length.

GUILLEMOT (*Uria aalge*)
Resident. It frequents rocky coasts, sea cliffs, stacks and islands, sometimes occurring in large colonies on the sheer, open cliff ledges, where they stand upright surveying all. The Guillemots in Britain have been divided into two species. The SOUTHERN GUILLEMOT (*Uria aalge albionis*) has dark chocolate-brown upper-parts, head and neck, and white underparts in Summer and brownish-grey upper-parts in Winter with white on the head and neck and a black line near the eye. The long, sharp-pointed bill is black and the legs and feet are yellowish brown, the plumage, feet and characteristic pose giving it a penguin-like appearance. The NORTHERN GUILLEMOT (*Uria aalge aalge*) is very similar in shape, plumage and habit to the SOUTHERN except it has very dark or black upper-parts and that it occurs chiefly in Scotland and the Scottish islands. Some of the NORTHERN GUILLEMOTS in Scotland and Shetland are "bridled" and have white ring around each eye and a white line runs back from this alon the sides of the head. These "bridled" guillemots are extremel uncommon in Southern Britain. Both species measure about 17 inche in length.

PUFFIN (*Fratercula arctica*) Fig. 13 (b)
Resident. It frequents cliffs, rocky coasts, turf-covered islands, sometim low and not precipitous, in numerous colonies, nesting in self-made animal burrows. The upper-parts are black, underparts white, the fa being greyish-white. Its legs and feet are orange-red in summer, winter yellow. The obvious aid to identification is its large, triangula shaped, parrot-like, laterally-flattened, grooved bill, which is bright coloured with red, yellow and grey-blue in summer, which gave it nautical nickname of the "Sea Parrot". During winter the bill is brov and orange as part of the outer bill is shed. It measures 12 inches length.

Fig. 13

(a) Razorbill (summer) *Alca torda* (b) Puffin (summer) *Fratercula arctica*

RAZORBILL (*Alca torda*) Fig. 13 (a)
Resident. It frequents the same habitats as the puffin and guillemot, sometimes breeding in large colonies. The upper-parts in summer are black with a black head and white underparts. There is a narrow white streak from the bill to the eye during summer, also a conspicuous, curved white bar on each black wing, particularly noticeable in flight. In winter he neck and sides of the head are white. The aid to identification is the hick, laterally flattened, axe-shaped black bill that has an almost vertical, prominent white line across its deepest, thickest part. The legs and feet are black. It measures 16 inches in length.

MISCELLANEOUS
In addition to the foregoing birds classed in families there are several ther birds which are either the sole species occurring in the British sles of a larger universal group of birds or the single member of a amily the remainder of which are entirely inland-based.

Fig. 14 Cormorant (winter) *Phalacrocorax carbo*

CORMORANT (*Phalacrocorax carbo*) Fig. 14
Resident. It commonly frequents estuaries, tidal rivers, rocky islands cliffs and any type of coast, also inland on lakes and reservoirs. Ofte it can be seen, with its wings displayed, outstretched as if drying them perched erect on a buoy, mooring post, pier structure, rock or simila objects projecting above the water. When swimming it does so low i the water with the bill pointing skywards and when pursuing fish i often dives to reappear a considerable distance away from the point o entry. It is powerful in flight with the long neck and short, webbed fee outstretched, normally just above the water surface. It is generally dar bronzy-brown and black, the plumage having a greeny-purplish shee The sides of the head and chin are white, there also being a white patc near the legs. The head is long and flat, with a reclining crest of feather the neck also being long, while the bill is hooked. It measures 36 inch in length.

GANNET or SOLAN GOOSE (*Sula bassana*)
Resident. Usually maritime, but in breeding season it frequents, i colonies, precipitous cliffs, islands and rock stacks. At first glance it cou also be mistaken for a goose or large gull, being entirely brilliant whi except for the buffish-cream head and neck, black marking from t

grey bill to behind the eye and black tips to the wings. The wings, however, are an aid to identification being very much longer, narrower and more pointed than the gulls, also the tail is pointed. Most spectacular is their flight when fishing. They dive towards the sea with their wings almost open, but just when it seems they must crash they check their flight momentarily and fold their wings at the last second to strike the water, disappearing in the spray, to impale a fish on the bill. It measures 36 inches in length.

RED-THROATED DIVER (*Gavia stellatus*)
Resident and winter visitor. It frequents offshore waters, sometimes close to the coast, also estuaries and large river mouths. In the breeding season the upper-parts and head are grey, the upper throat is also grey but the lower area is a dark chestnut-red, hence its name, the underparts being white. The back of the neck has white streak markings; the bill is upturned. In winter the upper-parts are brownish-grey with numerous white speckled markings. It continuously dives while being very low in the water. The Red-throated is the smallest of the three Divers, the others being the GREAT NORTHERN DIVER and BLACK-THROATED DIVER, which occur around our coastline during winter. It measures about 22 inches in length.

ROCK PIPIT (*Anthus spinoletta petrosus*)
Resident. It frequents rocky coasts, low cliffs, stony seashore beaches, mudflats and salt marshes, sometimes flat shores. It has dark, olive-brown upper-parts, the underparts being paler with numerous large buffish spots and streaks on the breast. The outer tail feathers are grey and prominent in flight. It is constantly active on the habitat running to seek out its prey, Sandhoppers, Insects, Flies on seaweed, etc. It measures about 6 inches in length.

SHAG (*Phalacrocorax aristotelis*]
Resident. It frequents in colonies rocky coasts, cliffs, stacks, and islands, and does not usually occur on tidal rivers or inland water. In summer it is a glossy dark greenish-black and has a conspicuous recurved crest on its head, there being no white in the plumage as its larger relative the cormorant. As well as being smaller the shag is a leaner bird, an aid to identification being the way it jumps upwards and forwards out of the water before diving beneath the surface to powerfully pursue fish. It measures 30 inches in length.

Chapter 3

Mammals

The Mammals [*Vertebrata*) which occur on the coast, estuary and seashore are few in number and only two, the Seals, are truly marine, but visit the land, the others being inland species. It may be that many hours will be spent before a wild animal is seen in these habitats. The site is important, it obviously being more likely that such mammal species will be seen in the remoter, quieter, less inhabited places, than close to developed areas or beaches busy with people. The foreshore of estuaries, the mudflats and creeks are more frequently visited by some of these mammals than the exposed coasts directly on the sea.

The COMMON OR BROWN RAT (*Rattus norvegicus*) frequents harbours and ports and their vicinity. It also commonly visits the remoter coastline and foreshore, more so at dusk or night, sometimes venturing as far down as the lower shore at low water, but more usually scrambling about the rocks or higher levels of other types of shore. Here it grubs among the debris cast ashore, for remains of human food, carrion and any other garbage it finds edible.

I have discovered the "tumps" or mole-hills ox the MOLE (*Talpa europaea*) thrown up in grazing pastures on the landward side but close to the sides of seawalls and the reed-lined drainage dykes of marshes bordering the North Kent mudflats and Medway Estuary.

The FOX (*Vulpes vulpes*) in its nocturnal search for food will visit the coast, estuary or seashore if within range of them, to prey on crabs and mussels and also during the breeding season to take eggs and young of various sea birds in their nesting sites. Reynard has also been credited with catching fish at the seashore in shallow water.

The COMMON LIZARD (*Lacerta vivipara*) occasionally occurs in suitably warm habitats within range of the sea, sometimes frequenting sand-hills and sandy, stony dunes, while its much rarer relative, the SAND LIZARD (*Lacerta agilis*) occurs in a few coastal sand-hills in northern England.

The RABBIT (*Oryctolagus cuniculus*) also occurs in dunes and gravelly sandy areas where there is a suitable food source or if the spot is near cultivated land.

Although chiefly thought of as an animal that occurs beside remote freshwater rivers, streams and pools the OTTER (*Lutra lutra*) sometimes travels overland at night during the summer months to visit estuaries and the open seashore. Here it lives in cliff caves or has its "holt" in a hole in a seawall embankment. Occasionally it can be seen among the rocks during the day. Some Scottish otters are known to spend as much time on the rocky coast as they do in their inland habitats. Nocturnally it hunts for crabs, mussels, small crustaceans, as well as searching in shallow water for various fish and it is said to dig in mud for eels, which it also takes when they are on migration.

Two SEAL species may sometimes be seen in habitats around our coastline.The hind, paddle-like limbs or flippers are used to push the seal at speed through the water, while the shorter front limbs are kept against the sides, but the limbs cannot support the weight of the body on land and so the seal progresses ungainly when out of the sea. On land they are clumsy and awkward, but as soon as they dive, into the water in swift pursuit of fish, with the ear openings in their fur and their nostrils closed, you can see immediately how well suited are the bodies of these sleek, streamlined animals to their aquatic way of life. They have to surface periodically to breathe but their bodies are constructed to conserve oxygen and so allow the seal to remain under water for a considerable time. All the watcher may see is the seal's whiskered muzzle and seemingly earless, dark limpid-eyed head jutting from the sea. To identify the two British species you need to have a side view, so that you can get a close-up sight of the nostrils. The female Seal gives birth to her one or two young, or "pups", at shore or offshore "rookeries" or "nursery" islands.

COMMON SEAL OR HARBOUR SEAL (*Phoca vitulina*) Fig. 15 (a) Common. It frequents estuaries, coast rocks, mudbanks and sandbanks, where it basks in the sun. It also regularly makes its way up tidal rivers and is a source of interest to the riverside population. This seal occurs chiefly on the east coast of England, eastern and western Scotland, the Hebrides, also in Orkney and Shetland and Northern Ireland, but does occur elsewhere when moving around our coasts. To identify it from the

Grey Seal in water, the Common Seal has a short muzzle and rounded head, the nostril slits being on an angle. Its colour is dark brownish-grey on the upper-parts and lighter on the underside, the body also having a varying number of dark spots. It feeds on various fish, molluscs, crabs and other crustaceans. In length the bull or male measures about 6 feet, the female or cow about 5 feet.

Fig. 15 (a) (top) Common Seal *Phoca vitulina*
(b) (bottom) Grey Seal *Halichoerus grypus*

GREY SEAL OR ATLANTIC SEAL (*Halichoerus grypus*) Fig. 15 (b] Common. It frequents rocky shores and islands and occurs on the west coast of England, Wales and Scotland, the Scottish islands, Orkney and Shetland and the Farne Islands off the Northumberland coast and the east coast of England in The Wash area. The head is longer than the Common Seal and has a flattened top, the nostril slits being nearly vertical. It also varies considerably in colour. The male or bull is very dark brown or black varying to greyish, with paler areas and spots; the female or cow varies from light brown to greyish, the underside being paler with darker spots. It feeds on fish, squid, cuttlefish, crabs and other crustaceans. In length the male or bull measures up to 8 feet, sometimes more, the female less, about 7 feet.

Chapter 4

Fish

Little need be said as an introduction to the Fish (*Pisces*) as their basic anatomy is well-known, except that there are variations among the species in the shape and length of the body, in the shape, number and position of the fins, shape of the gill covers, positioning of the mouth and eyes, most fish being covered in scales, but some are scaleless.

A fish is a Vertebrate, having a movable "backbone" of vertebrae and spinal column. The body shape is streamlined for its life in water, the head and body being joined without a neck. The fish breathes by use of its gills at the rear of the head. It moves by muscular action of its body, using its tail and fins to steer forwards, upwards and downwards and balance itself. Most fish begin life as an egg, the viviparous birth of living young being rare in this type of animal. The eggs either float in the upper sea among the plankton or sink to the bottom. Some eggs, however, are laid in masses in the usual habitat or in a seaweed nest. After hatching the young fish fry develop into adults, taking a varying length of time to do so, according to the species. There is variation in the breeding cycle of those species of fish, the Salmon, Shad and Sea Trout, which migrate from the sea and pass through estuaries to spawn in fresh water rivers. These have not been included in this book, nor those species which are only occasional visitors from the sea to our coast, estuaries and river mouths, such as the Sturgeon, Monk or Angler-fish, because we are primarily concerned with coastal species or those frequently to be met in the habitats covered by the book.

THE BLENNIES (*Blenniidae*)

Some of these fishes commonly occur around our coastline, between the tide marks, while other species inhabit only certain areas. They also frequent varied pools and some are found in brackish water and estuaries. Each has an elongated, cylindrical body, with a rather large head and eyes and large, rounded pectoral fins on its sides, and they are not particularly attractive in appearance. An aid to identification is the dorsal fin which is continuous or nearly continuous along almost the complete length of the back.

BUTTERFISH OR GUNNELL (*Centronotus gunnellus*) Fig. 16 (a)
Common. It frequents various coasts in shallow water, particularly rocky and stony shores if these are seaweed-strewn. It prefers to hide under these situations when the tide recedes rather than in open pools. Its shape is elongated, with rather a similarity to an eel, laterally flattened and covered with very tiny scales. In colour it is brownish-buff, reddish-brown or a greyish-brown, the fins being mustard-yellow, reddish-brown or dark grey. An aid to identification is the ten to twelve black or dark spots, with a white ring around each, along the bottom of the continuous dorsal fin on the back. The eggs are laid in small masses in rock cavities, empty shells, and holes made by rock-boring bivalves. The parent fish guard the eggs until they hatch by coiling their body around them, to protect them from predators and to prevent the eggs from drifting and being scattered. Parental care of their eggs by fish is rare, especially by both parents. Usually only the male shows any interest in the eggs. It gets its name Butterfish from the slimy, slippery body which makes it extremely difficult to grasp. It measures up to 11 inches in length, but about 5 inches is usual.

BUTTERFLY BLENNY (*Blennius ocellaris*) Fig. 16 (b)
Locally common on south-west of England coasts, rare elsewhere. It frequents rocky shores with plenty of seaweeds. It is grey in colour with several lighter vertical marks, and has a short, blunt head with twc small tentacles over the eyes. An aid to identification is the large elevated, wing-like dorsal fin, half of which is higher at the front enc than the hind part. The back bears a conspicuous large, rounded white ringed black "eye-spot". It measures up to 7 inches.

COMMON BLENNY, SMOOTH BLENNY OR SHANNY (*Blenniu pholis*) Fig. 16 (c)
Very common. It frequents all types of coast, but particularly shallov rock pools with plenty of stones and seaweeds. Its colours are variabl and difficult to see in this habitat because the greenish-olive, dark gree or yellow body is blotched or banded with black or dark spots or patche which help it to merge into its background. The fins are a yellowish o greenish colour, also spotted with black. During the breeding season i

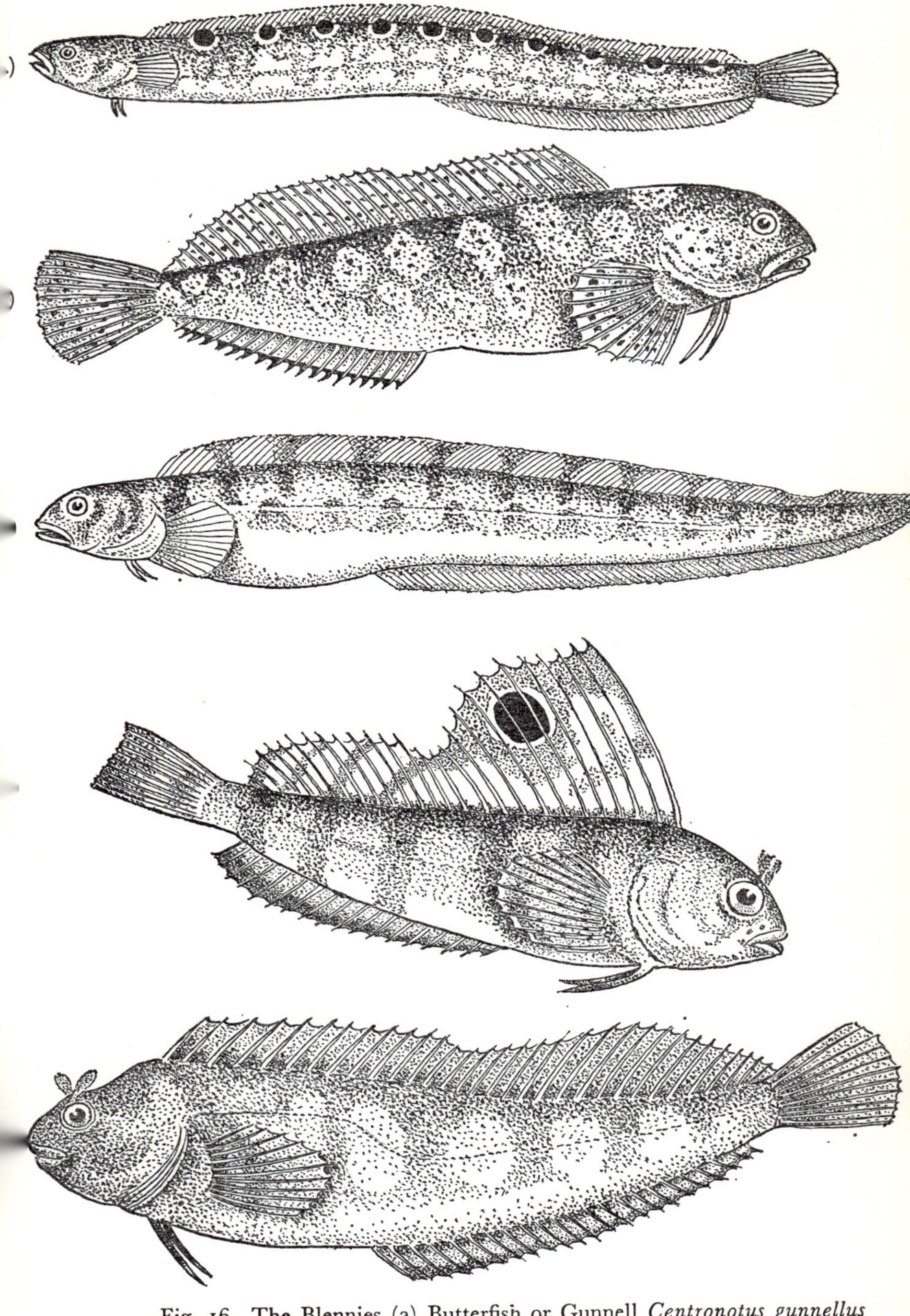

Fig. 16 The Blennies (a) Butterfish or Gunnell *Centronotus gunnellus*
(b) Butterfly Blenny *Blennius ocellaris*
(c) Common Blenny or Shanny *Blennius pholis*
(d) Tompot Blenny *Blennius gattorugine*
(e) Viviparous Blenny or Eel-Pout *Zoarces viviparus*

mid-summer the male changes to a uniform sooty-black, while his prominent lips remain a whitish colour. In this coloration he guards the eggs until they hatch. It is claimed that this fish deliberately leaves the water to bask in the sunshine upon rocks and weed. The body has no scales, hence its name of Smooth Blenny. It measures up to 5 inches.

TOMPOT BLENNY OR GATTORUGINE (*Blennius gattorugine*) Fig. 16 [d)
Locally common on south and south-west coasts of England, otherwise rare, though it frequents deeper water than other Blennies. It has been trapped in crab pots. It is dull greyish-brown or olive in colour, with olive-yellowish fins. An aid to identification is the conspicuous pair of fringed tentacles pointing upward close above the eyes. It measures up to 9 inches.

VIVIPAROUS BLENNY OR EEL-POUT (*Zoarces viviparus*) Fig. 16 (e)
Common on north and east coasts, occasional elsewhere. It frequents weed-strewn sandy shore. During the summer it migrates into deep water away from the habitat. It has a long, tapering body, with a low, long dorsal fin, extending the full length of the back which has a short-spined depression near the tail. There is a long anal fin on the underside, but it has no tail fin. The body is pale brownish-green or olive with lighter, greenish or brownish markings and curving patches on the upper body and dorsal fin, the fins being greenish-white. It gets its name from the way in which the eggs are fertilised by sperm from the male, develop and hatch inside the ovary of the female, the young remaining there for three months until finally born about four months after fertilisation, early in the year, measuring $1\frac{1}{2}$ inches in length. This method of development is rare among fish. It measures up to 24 inches but 6 to 14 inches is more usual.

THE ARMED BULLHEADS (*Agonidae*)

The Armed Bullheads are members of a different family of fish and are not related to the Bullheads or Sea-Scorpions (*Cottidae*) despite their similar names. Only one species is found in British inshore water.

ARMED BULLHEAD OR POGGE (*Agonus cataphractus*) Fig. 17
Locally common on south-east and north-east coasts. it frequents estuaries, river mouths and the vicinity of harbours and is also sometimes caught in shrimp nets. An ugly fish, with an elongated body that narrows from the central fins to the tail, and flat head, covered with bony

plates. The upturned snout is spined, while the lower jaw is bearded. It is pale yellow and grey, with darker grey-brown upper surface bands and spots, the fins being greyish-yellow. It measures up to 5 inches.

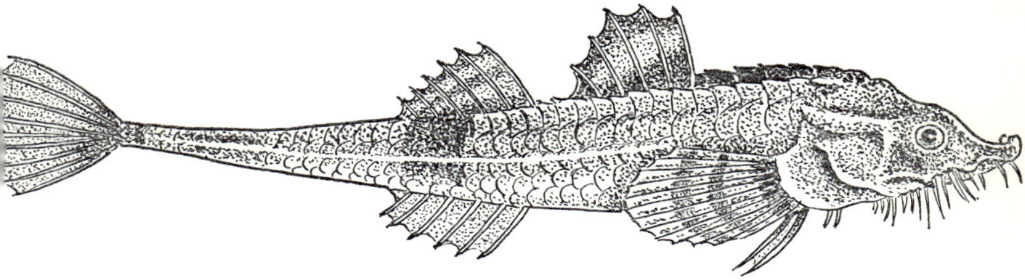

Fig. 17 Armed Bullhead or Pogge *Agonus cataphractus*

THE COD FAMILY (*Gadidae*)

A brief mention can be made of several fish in the Cod Family as they spend the early part of their life either close inshore or in rock pools and can be found there.

COD (*Gadus callarias*)

Spawns from January to March in the North Sea at a depth of 20 to 50 fathoms. The fry drift into shallow coastal water, particularly on the north and east coasts. In the first year they reach 7 to 8 inches in length if conditions are favourable. The young are called Codling. The familiar Cod is brown or olive-greenish on its upper surface with dark marbled spots, the underside being silvery-white. On the lower jaw it has a long barbel.

IVE-BEARDED ROCKLING (*Onos mustela*) Fig. 18

Common on all types of coast, particularly in sand or rock pools, and also frequents brackish water. The young lurk under seaweeds, stones nd shells in pools. It is greyish-reddish or olive colour on the upper-arts, with lighter hue on the underside, the fins being dark with orange arking near their base. An aid to identification is the four projecting arbels on the snout's upper lip and one on its chin. Unlike the Three-earded Rockling it has no spots on its body. It measures up to 12 inches.

HREE-BEARDED ROCKLING (*Onos mediterraneus*)

ommon on south and west coasts of England, occasional elsewhere. frequents similar habitats to the Five-Bearded Rockling. It is greyish-ddish on the upper-sides to whitish-grey on the underside, but some

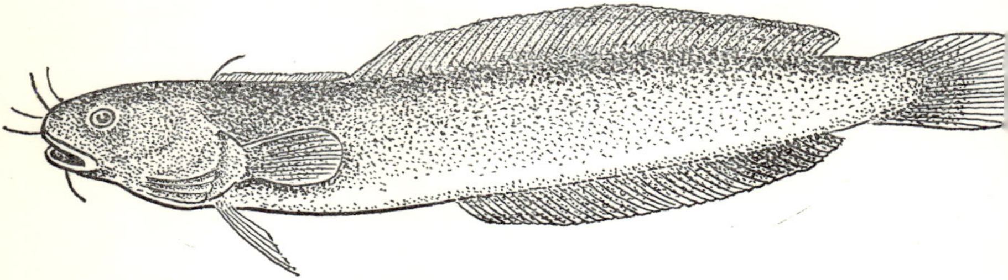

Fig. 18 Five-bearded Rockling *Onos mustela*

examples may have round brown or black spots. The aid to identification is the two barbels on the upper lip and one barbel on the chin. It measures up to 18 inches.

WHITING (*Gadus merlangus*)
Common, particularly on the east coast in shallow water, young examples frequenting estuaries and sandy bays and are commonly caugh inshore in the nets of shrimpers, in mid-summer. It is similar to the Pollack in shape, but lighter brownish in colour, with silvery sides anc a prominent black spot bordering the pectoral fin, the fin nearest tc the gills. The lower jaw does not project so far beyond the upper jaw as the Pollack and it has no barbel on the chin.

THE EELS (*Anguillidae*)
The Eels have elongated, round bodies in which the head and tai are combined to give the fish a snake-like appearance. The dorsal, ana and tail fins are continuously combined, the pelvic fins being absent They are extremely active and their supple, slippery body is difficult t grasp. The saga of how eels are born from eggs laid in the vicinity c the Sargasso Sea, then, as larvae, cross the Atlantic Ocean, the journe taking three years, change into elvers and mature, either by ascendin Europe's freshwater rivers or living in saltwater around our coast depending on the species, is a fascinating life story that has been full told elsewhere.

COMMON EEL (*Anguilla anguilla*) Fig. 19 (a)
Common. It spends most of its life in freshwater rivers, water-meado dykes, lakes and ponds, the latter, although land-locked, being reache overland through wet grass and soil on dewy mornings. The glistenin

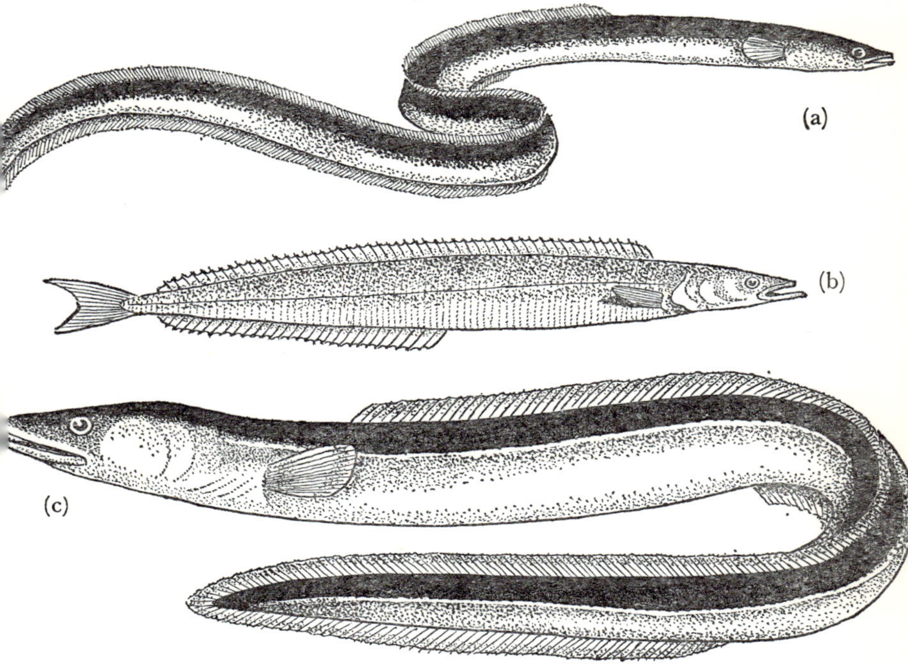

Fig. 19 The Eels (a) Common Eel *Anguilla anguilla*
(b) Lesser Sand Eel *Ammodytes lancea*
(c) Conger Eel *Conger conger*

eel-like young elvers, up to 6 inches in length, are found on the coast, particularly where there is access to streams, water-courses, dykes, also in estuaries in springtime, making their way up-river where they will feed and develop. On reaching maturity, between six and ten years of age, they descend the river, estuary, etc. to the sea, in late summer, en route to their spawning ground. Much confusion has arisen over its description and whether or not there are several species of Common Eel. The "Silver Eel" or "Sharp-nosed Eel" is only the Common Eel in its breeding dress during the autumn when on migration; the "Yellow Eel" is the coloration of yellowish-brown above, off-white below, of the Common Eel. The "Frog-mouthed Eel" or "Broad-nosed Eel" is the name given to Common Eels which have larger than normal development of their jaw muscles; such eels, usually females, are thought to become this way due to being more voracious than usual. There is, however, considerable variation in coloration; most common is that the males are a bronzy-yellow or green on the upper-parts, whitish below, the females olive-green or yellowish-brown on the upper-parts and yellowish below, both sexes changing to migratory coloration. Eels feed chiefly at dusk or night, hiding by day under rocks, stones, in crevices and holes in the river or dyke bank. Young eels feed on marine worms, crustacea; adult eels are greedy, taking small water birds, voles, shrews, frogs, other fish and various marine creatures. The adult male measures up to 20 inches, the female up to 36 inches although specimens considerably larger have been caught.

CONGER EEL (*Conger conger*) Fig. 19 (c)
Common on south and west coasts of England, occasional elsewhere. Small examples, up to 24 inches long, occur in shore pools, and much larger specimens around the coasts, sometimes lurking among mud and under stones on the lower shore. This eel does not venture into fresh-water rivers, although other parts of its life cycle are similar to the Common Eel and it also migrates to spawn in deep water in the Atlantic Ocean. The young Conger Eel is pale, pinkish in colour, but as it grows it becomes greyish on the upper-parts and off-white on the undersides. It has no scales. The jaws are well-armed with rows of teeth.

THE SAND-EELS (*Ammodytidae*)
Although their smooth bodies are eel-like at first glance Sand-Eels are not related to the true eels and have several differences. The Sand-Eel's body is also elongated but compressed not round, the complete dorsal fin along the back and the shorter anal fin on the underside are both separate from the tail. They sometimes occur inshore in huge shoals in the shallow water of sandy bays during spring and summer. A shoal may be indicated by numbers of diving sea birds taking the opportunity of securing an easy meal. Other fish and porpoises also feed on them. Although quick in movement the Sand-Eels use their large, spoon-shaped lower jaw to dig into the sand and bury themselves to try to escape and also remain there during low tide. They feed on crustacea, other fish, even their own young.

GREATER SAND-EEL (*Ammodytes lanceolatus*)
Fairly common in suitable localities. It is yellowish-green or bluish-yellow on the upper-parts and pale silvery-white on the sides, with a darker area near the upper jaw. The spoon-shaped lower jaw projects beyond the upper jaw. The dorsal fin on the back starts behind the pectoral fins. It measures up to 12 inches in length.

LESSER SAND-EEL (*Ammodytes lancea*) Fig. 19 (b)
Common in the same localities as the Greater Sand-Eel. It is bluish-green or yellowish-green on the upper-parts, silvery white below and on the sides. Two differences as aids to identification are that in the Lesser the dorsal fin on the back starts in the centre of the pectoral fin and the snout tapers more distinctly to the mouth than in the Greater, the lower jaw being very protractile. It measures up to 8 inches in length.

THE FLATFISH (*Pleuronectidae* and *Soleidae*)
Several of the familiar flatfish spend part of their early life as shore fishes or entire life in estuaries, mudflats and sandy bays. The upper surface of the fish is coloured to resemble the bed of sand, mud, etc., on or in which they live or bury themselves to hide from predators. The underside, which is hidden while at rest, is white or a pale colour. They usually spawn in deep water, the eggs and larvae drifting shorewards into shallower water. Here the amazing change from a normal fish fry to the flatfish takes place, the eye on one side moving round the head so that both eyes are on the same side of the body. This is obviously necessary to a fish that has one side of its body and head downwards upon the mud, sand, etc., and the mouth pointing upwards. Dabs, Plaice, Flounders, Sole, lie on their left side, with both eyes on the right side, the left or underside being pale coloured, while Brill and Turbot lie on their right side, consequently their eyes are positioned on the left side, the right or underside being light coloured.

BRILL (*Rhombus laevis*)
Fairly common, more so on southern coasts, on muddy, sandy bottoms, the young coming close inshore in shallow water in bays. Oval in shape, the left or upper side is a brownish-green with large and small dark brown and paler spots and blotches; the underside is white; it also has smooth scales. It measures up to 3 inches in length when caught inshore.

DAB (*Pleuronectes limanda*)
Common. It frequents inshore waters, especially shallow, sandy bays. Young dabs, with Plaice, are frequently caught in shrimpers' nets, particularly during September and October. The right or upper-side is brownish with a greenish hue and sometimes has dull yellow spots, the spots being arranged similar to the Plaice, but not so large or distinct; the underside is a light brownish-grey. The skin, due to the spiny edges

of the scales, feels rough. On its upper-side the Dab has a clear curve in its lateral line near the pectoral fin, which is an aid to identification. It is rounded-oval in shape and measures up to 10 inches in length.

FLOUNDER (*Pleuronectes flesus*) Fig. 20
Very common. It frequents all types of coast, occurring particularly in muddy estuaries and muddy or sandy bays, also ascends rivers to inhabit brackish water and sometimes fresh water. In the Medway estuary creeks fishermen, at low tide, on the mudflats, catch Flounders by the method known as "griping" for them in the streams where there is still a trickle of water. The fisherman gropes carefully in the mud until a Flounder is detected lying on the bottom and then, with one hand behind the fish's head, the other hand over the tail, the Flounder is grasped and quickly hauled through the mud and water to be dropped in an open sack tied round the fisherman's waist. The flesh of the Flounder is considered by some people inferior to the Plaice and other flatfish, but I do not agree. Although it may spend much or all of its life in rivers and estuaries it returns to the sea to spawn. This seaward migration usually takes place in March and April, although in some mild or cold seasons spawning may take place in February or continue into May. The left or underside of the rounded-oval-shaped Flounder is pearly-white and an aid to identifying it from other flatfish, although examples are also found with areas of light brown blotches and the white skin sometimes appears to be discoloured by oil and other pollution

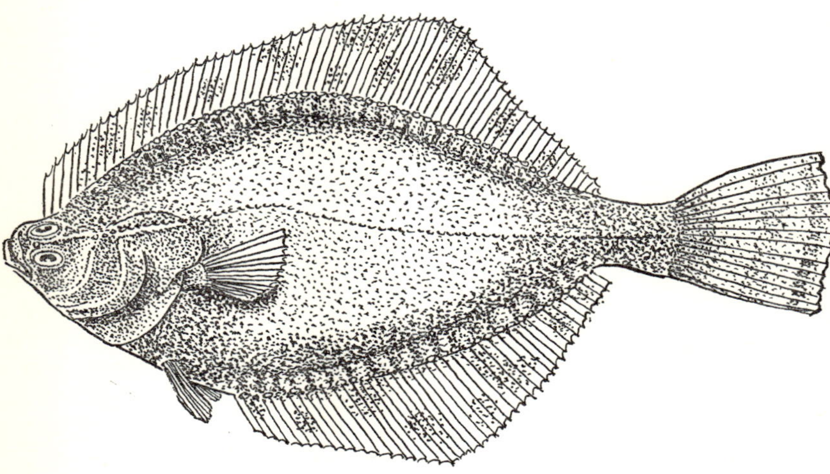

Fig. 20 Flounder *Pleuronectes flesus*

when the fish lies in mud containing these. The right or upper side varies according to its surrounding background. In mud it is generally a dark brown, sometimes greenish-black or almost black, but it can also be olive-brownish or variously mottled to match shingle, or brownish-yellow on sandy shores. The lateral line on the upper-side has small bony protuberances. The Flounder is attacked by several types of parasite, one of which causes round, whitish swellings on the skin exterior. It measures up to 18 inches.

PLAICE (*Pleuronectes platessa*)
Very common. It frequents all coasts. Young examples, several inches in length, are captured, with Dabs, in shrimpers' nets inshore. The right or upper-side is dark greenish-brown, with numerous large orange-red spots, sometimes surrounded by a white circle. In immature fish the white is absent. The left or underside is greyish-white. It has bony knobs on the head between the eyes. The head is larger in proportion to its round-oval body than is the case with other flatfish; the body is also broader in proportion with its length. The lateral line on the upper-side is almost straight. It measures up to 18 inches, occasionally more.

SOLE (*Solea solea*)
Common. It frequents all coasts and may be found in pools in shallow sandy localities during the summer months. The right or upper-side is a brownish or greyish-green, with darker markings and sometimes lighter spots; the underside is white. There is a dark marking at one end of the pectoral fin, the body being long-oval and not rounded-oval. It usually measures up to 18 inches, but is occasionally longer.

THE GOBIES, SEA-SNAILS, LUMPSUCKER & SUCKER FISH

The Gobies (*Gobiidae*) are among our smallest fish species. They frequent pools on rocky coasts and estuaries. Here they use their two modified pelvic fins underneath their body as a fan-like sucker to attach themselves to the rocks. In these pools most of the Gobies, which are usually found in shoals, live at the bottom and hide themselves among the stones and empty shells or sometimes lie on the sand itself, the exception being the Spotted Goby. Because of this they are difficult to see, so patience and quickness of the eye are required to achieve this. There are two separate dorsal fins on the back but an aid to identification of the Gobies is their prominently large eyes which almost touch in some of the species. To distinguish between the species the colour and shape of the dorsal fins is a guide. There are ten species of Goby around our coasts, but only the common examples likely to be found are described.

BLACK GOBY (*Gobius niger*) Fig. 21 (b)
Fairly common. It frequents muddy, rocky and sandy shores and is also found in brackish water and in estuaries. Its coloration varies from yellow-ochre, brownish or blackish-grey, the sides being speckled and marked with lighter and darker areas. The front dorsal fin has a dark top edge and six spines. It measures up to 5 inches.

COMMON GOBY OR SAND GOBY (*Gobius minutus*) Fig. 21 (a)
Very common. It frequents sandy pools and muddy, brackish estuaries in large numbers. Its colour is variable, according to the habitat. On shallow sandy shores it matches its background, being a dark yellowish-brown on its upper side with several darker freckle or blotch markings on the sides; in muddy estuaries it varies to a darkish grey. An aid to identification is the single dark blue or black spot conspicuous on the rear end of the front dorsal fin. Its length also depends on the habitat.

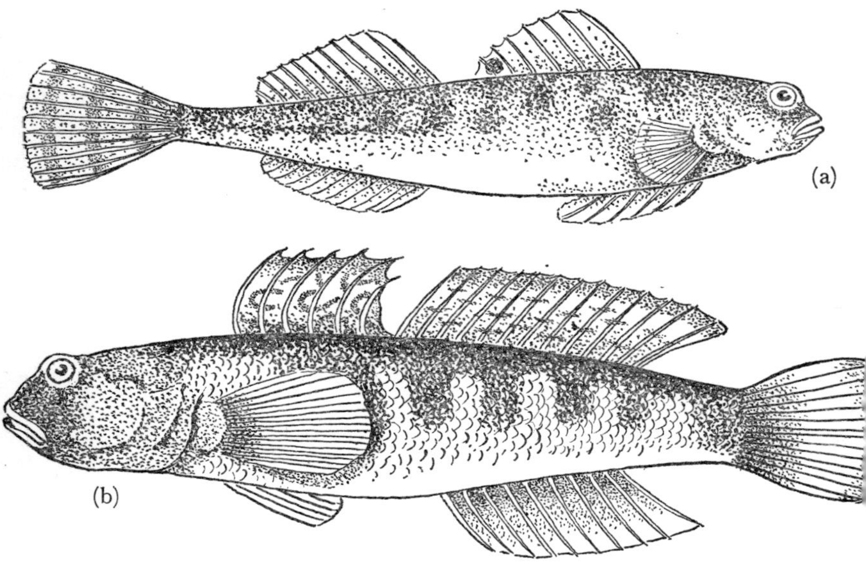

Fig. 21 (a) Common Goby or Sand Goby *Gobius minutus*
(b) Black Goby *Gobius niger*

In brackish estuaries it may measure up to 2½ inches, but in salt-water shore pools may be up to 4 inches.

PAINTED GOBY (*Gobius pictus*)
Fairly common, more so on south and west coasts of England. It usually frequents pools on coarse sandy shores in large numbers. Its coloration is pale greyish-yellow or pale brown-grey, with lighter markings, and two dark markings below the eyes. It has longitudinal reddish bands on its pale dorsal fins and black spots at the base of each fin are an aid to identification and give it its name. It measures up to 2½ inches.

SPOTTED GOBY (*Gobius ruthensparri*)
Common. It frequents rock pools, but does not lurk on the bottom like other Gobies. Instead it swims freely about in shoals among seaweeds. Its coloration varies from a yellowish-brown to olive-brown, with several paler, irregular, saddle-like marks close to the dorsal fins. The seven-spined front dorsal fin has three horizontal bands of reddish or pink with areas of yellowish-white or white; the hind dorsal fin has three or four similar reddish or pink bands. An aid to identification is a large black spot on the body near the head and the central base of the tail. It measures up to 4 inches.

The Sea-Snails or Smooth Suckers (*Liparidae*] are not snails at all but small fish, yellow in colour, measuring several inches in length. They are most frequently found in rock pools and on rocky, seaweed-covered shores. The smooth head is round, broad and blunt, while the body is sub-cylindrical with a semi-loose, scale-less, smooth skin and a single dorsal fin. The pelvic fins are modified for use as a sucker.

COMMON SEA-SNAIL OR COMMON SUCKER (*Liparis liparis*)
Fairly common. It frequents rock pools, lurking under the seaweeds, also in shallow water during the summer when they are caught in shrimpers' nets. During the winter it migrates to deeper water where it spawns from January to March, laying its eggs in masses on seaweeds. Its coloration is light greyish to brownish, the head and body being marked with irregular, horizontal dark bands or lines, the yellow tail having darker markings. It measures up to 6 inches.

MONTAGU'S SEA-SNAIL OR MONTAGU'S SUCKER (*Liparis montagui*) Fig. 22
Common. It frequents similar habitats to the Common Sea-Snail, but

more so in estuaries. It spawns from January to March also laying its eggs on seaweeds. It is reddish-purple to yellowish-brown in coloration, resembling the Common Sea-Snail in appearance, but instead of lines has numerous small dark spots on the body. The yellowish fins are also black spotted. It meaures up to 4 inches.

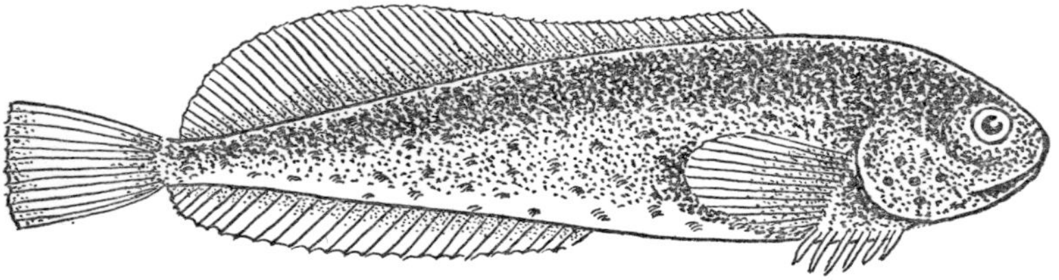

Fig. 22 Montagu's Sea Snail *Liparis montagui*

LUMPSUCKER OR SEA-HEN (*Cyclopterus lumpus*) Fig. 23
Fairly common, more so on northern coasts. The adult Lumpsuckers (*Cyclopteridae*) frequent our coasts in spring when they come ashore to spawn. The young are often numerous in rock pools during summer. Easily identified by its curious, remarkable appearance. It has a scale-less, thick, lumpy, rounded, deep body, with four lines of tubercles on each side. There is no front dorsal fin, only a row of lumps in its place. The head is large and rounded, with a short snout. Beneath the body there is a conspicuous and efficient adhesive, sucker-disc, similar to that of the Gobies. The Lumpsucker, using the sucker, attaches itself with great firmness to rocks and other objects and is difficult to remove, sometimes the object also being lifted with the fish. It is lazy and sluggish and is said to fix itself to an object and then waits for food prey, crustacea, etc. to pass close by before it will feed. In colour both sexes are usually a blackish-green, though the male, particularly during the breeding season, has a brick-red coloration on the sides and underside, the female then being dark bluish or blackish with yellowish-white on the underside. The male faithfully guards the egg mass, keeping a flow of water over it with his fins. Diligently he will remain, sometimes fixed to a stone or rock close to the eggs as the tide recedes and is then a victim of gulls and other birds. In rough weather or at low spring tides the sluggish Lumpsuckers are often cast up or stranded on the shore where they die. Examples have reached up to 24 inches in length, but it is more frequently found up to 8 inches.

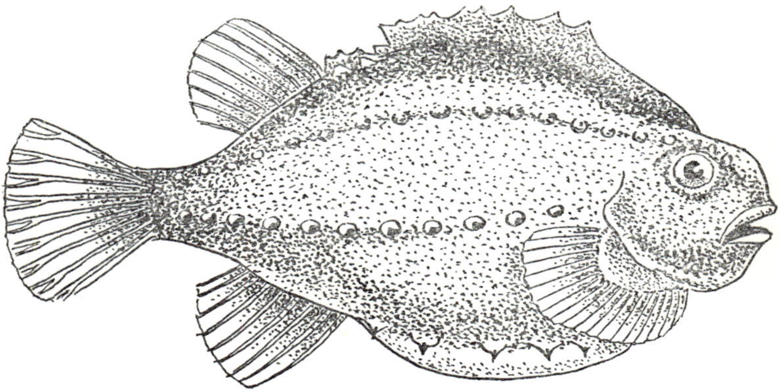

Fig. 23 Lumpsucker *Cyclopterus lumpus*

The Sucker Fish or Cling-Fish (*Gobiescidae*) are small scale-less fishes which as the name indicates, are equipped with a large, adhesive sucker-disc formed from the pelvic fins and a modified gristly bone, the disc being about one-third of the body length. An aid to identification is that the dorsal and anal fins are both sited at the rear end of the body. They frequent rocky coasts, lurking under stones in pools during summer, most frequently occurring in the south-west of England. During the winter they migrate into deep warmer water. Spawning takes place in June or July, the eggs being laid on stones and seaweed or in empty shells, guarded by one or both parents.

CORNISH SUCKER (*Lepadogaster gouanii*) Fig. 24
Common on west coasts, uncommon elsewhere. It is reddish or purplish, occasionally green or brown, with lighter undersides and some yellow lines between the eyes. On the "shoulders" it has two dark, round spots, each surrounded by a whitish ring. The dorsal fin and anal fin are combined with the tail fin. It is the largest of the Sucker Fish, measuring up to 4 inches.

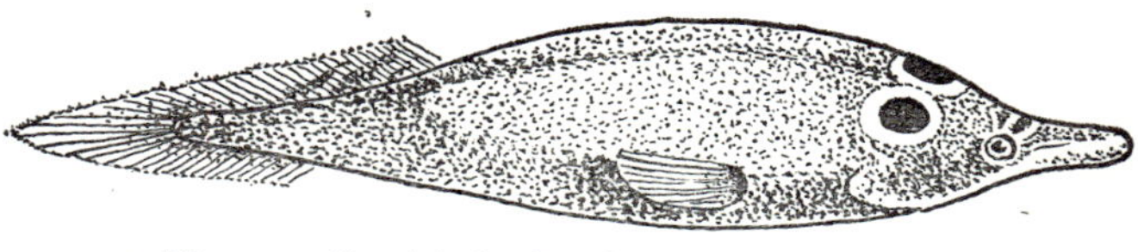

Fig. 24 Cornish Sucker *Lepadogaster gouanii*

TWO-SPOTTED SUCKER (*Lepadogaster bimaculatus*)
Common. The orange-reddish body has dark "eye spots", surrounded by a lighter colour, on the sides near the pectoral fins, and red V-shaped lines from near the eyes. The dorsal and anal fins are separated from the tail fin.

THE PIPE-FISH (*Syngnathidae*)
This family are aptly named, each having a long, thin, tapering, pipe-like body, small fins, narrow extended head and tubular mouth. This body shape is excellent camouflage for these fish, helping them to hide among moving seaweed stems. They move through the water by rapidly vibrating their brief, separate dorsal fin, although they are weak swimmers and often at the mercy of the currents. On the underside of some male Pipe-Fish species there is a pouch formed from a fold of the skin, in which the eggs are incubated after being laid there by the female. After hatching, the young stay close to the male and if danger threatens they return to the pouch for protection. The Pipe-Fish occur close inshore on our coasts, also entering estuaries and living in brackish water and may even move into fresh river water where it flows into the estuary. They are related to the well-known Sea Horse.

BROAD-NOSED PIPE-FISH (*Syngnathus typhle*)
Fairly common. It frequents inshore waters, in shallow, seaweed-covered habitats, spawning in May and June when it migrates into deeper water. The male carries the eggs in his brood pouch until they hatch. It is dark brown with paler spots on the underside. The snout is large but shorter and deeper than in the Greater Pipe-Fish. The body, which is fattest behind the gills, has ridges which run lengthwise. It measures up to 12 inches.

GREATER PIPE-FISH, NEEDLE-FISH OR HORN-EEL (*Syngnathus acus*) Fig. 25 (a)
Fairly common. It frequents particularly areas where eel-grass (*Zostera*) grows and is sometimes captured in shrimpers' trawls. It spawns during Spring and Summer, the eggs also being carried in the male's brood pouch. It is brown on the upper parts with lighter brown below, also has darker band markings along the sides. The snout is very long and thin, while the body, also with ridges running lengthwise, has a paunchy appearance being fatter in its middle, between the gills and the dorsal fin, before tapering to the tail from the dorsal fin. It measures up to 18 inches.

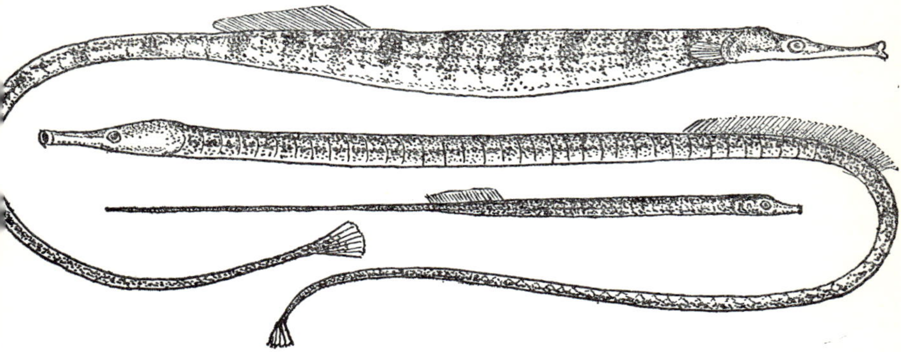

Fig. 25 The Pipe-fish (a) Greater Pipe-fish *Syngnathus acus*
(b) Snake Pipe-fish *Entelurus aequoreus*
(c) Worm Pipe-fish *Nerophis lumbriciformis*

SNAKE PIPE-FISH (*Entelurus aequoreus*) Fig. 25 (b)
Fairly common. It occurs around our entire coasts and is sometimes captured while still clinging by its tail to flotsam, seaweeds, mooring ropes, etc. The eggs are not contained in the male's pouch, but are glued in rows by the female to the abdomen of the male, who carries them about in this way until they hatch. It is olive-green, with numerous thin, bluish stripes on the sides, reaching to the rear edge of the yellowish dorsal fin. The body is extremely long and narrow, round and smooth, not ridged. The snout is elongated with a purplish stripe beginning at he eyes. Sometimes the tail fin is missing in large examples or only very small. This is the largest British Pipe-Fish measuring up to 30 inches, but 15 to 20 inches is average.

WORM PIPE-FISH (*Nerophis lumbriciformis*) Fig. 25 (c)
Common, more so on the west coast. It lurks among brown seaweeds which it matches ideally. It spawns in early summer and like the Snake Pipe-Fish the female attaches the eggs in rows to the abdomen of the male who carries them until they hatch. It is olive-brown with tiny, paler lines and dots on the underside. The body is round, smooth, tapering and very thin, with a short snout and has no pectoral fins and tail fin. It feels stiff and hard like the seaweed stems it resembles. The smallest British Pipe-Fish measures up to 6 inches.

THE SEA-SCORPIONS OR BULLHEADS (*Cottidae*) AND TING-FISH OR WEEVERS (*Trachinidae*)
The Sea-Scorpions or Bullheads are an ugly species and, like the land

Scorpion, are armed to repel would-be predators. Despite their name they are not related to the Armed Bullheads. The Sea-Scorpions (*Cottidae*) have sharp, projecting spines for this purpose of repelling. When picked up by sea birds and other fish the Sea Scorpions immediately raise their spiny gill covers to broaden their flattish head. This sometimes prevents them from being swallowed by the predator who, finding the intended meal is uncomfortable and likely to stick in its throat, has to release it. This species has large depressed heads rounded in front, which gave them their name of Bullheads. The spiny, scaleless body, with two dorsal fins and other large fins, is irregular in shape. With their mottled and banded coloration they are able to blend into the background and hide among stones and seaweeds, darting forward only to sieze a crustacean which is consumed with one gulp of the large mouth. Otherwise they are inactive swimmers and tend to lurk about in the sanctuary of rocks. Both spawn during the winter and spring among rocks.

SHORT-SPINED SEA-SCORPION, FATHER LASHER, OR BULLHEAD (*Cottus scorpius*) Fig. 26

Common, particularly on north and west coasts. It frequents rock pool and shallow water in summer, among seaweed, and is occasionall caught in shrimpers' nets. It varies in colour, brownish-grey or yellowis marbling, with darker bands and spots; the underside is patterned wit reddish-brown, greyish-white or yellowish-brown, the fins being brown yellow with red spots. An aid to identification is a spine above the ey and the four spines on the gill covers. It can produce a grunting soun

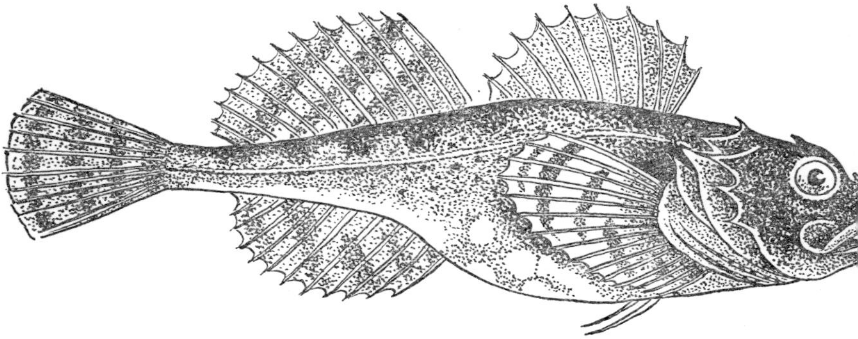

Fig. 26 Short-spined Sea Scorpion or Father-Lasher *Cottus scorpius*

A pair of Kittiwakes adding their voices to the din of a cliff-face nesting colony.

A Shag upon its untidy, often smelly, nest on a rock ledge.

An Oystercatcher *Haematopus ostralegus* in flight.

The teeming activity of a Kittiwake colony.

when removed from its habitat. It usually measures up to 10 inches, but larger examples have been caught.

LONG-SPINED SEA-SCORPION OR COBBLER (*Cottus bubalis*)
Common, particularly on west and north coasts. It frequents rock pools and shallow water inshore during summer and moves, like the Short-Spined Sea-Scorpion, into deeper water to breed in winter. It is also variable in colour, sometimes resembling the Short-Spined, but occasionally more brilliantly coloured in bright or deep red, yellows and greens, particularly during the spawning period. An aid to identification are the five spines on the gill covers, one being particularly larger and longer. It measures up to 12 inches, but 8 inches is usual.

The Weevers or Sting-Fish (*Trachinidae*) are dangerous fish to capture. The spines on the front or anterior dorsal fin, pelvic fins and gill covers can inject a poison that causes painful swellings. Dead examples can also do this. These fish lie in or on sandy habitats where shrimps are numerous, sometimes being themselves caught in the shrimpers' nets. If you handle or tread barefoot on one and are aware you have been stung you must immediately obtain medical aid.

GREATER WEEVER OR STING-FISH (*Trachinus draco*)
Fairly common. It frequents sandy coasts and shores in shallow water, more so in summer, and also occurs at a further distance from the shore in deeper water. It also lies in the sand and has poison-injecting spines, but is not so dangerous as the Lesser Weever. It spawns during the summer. It is similar in appearance to the Lesser Weever, being greyish or yellowish-grey, darker on the upper-parts, sometimes having thin brown markings on the sides. Part only of the front dorsal fin is black, the other portion being a lighter colour. An aid to identification are the two small spines sited near each eye. It measures up to 16 inches, but to 10 inches is usual.

LESSER WEEVER OR STING-FISH (*Trachinus vipera*) Fig. 27
Fairly common, more so on south-west coasts. It frequents shallow water and sandy shores, where it lies in the sand with only part of its head, the eyes, upward-pointed mouth and lips and tips of the erect spines of dorsal fin above the surface. It is frequently caught in shrimp nets, as it feeds mainly on shrimps and is a hazard to shrimpers. It spawns from May to September in shallow coastal water. It is greyish-sandy brown with paler yellow on the sides. The anterior or front of the two dorsal

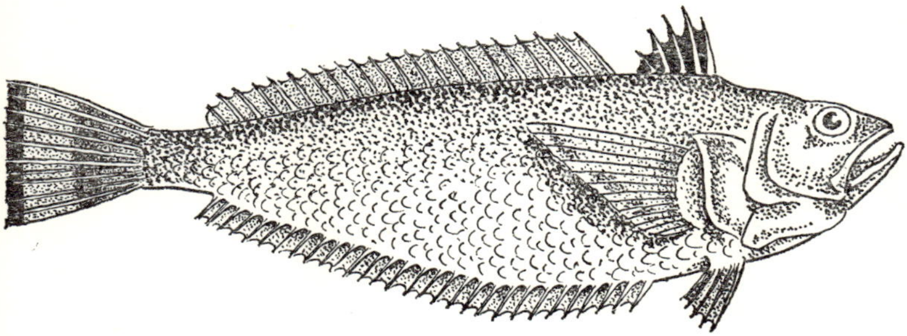

Fig. 27 Lesser Weever or Sting-fish *Trachinus vipera*

fins is conspicuously black and bears five spines. There are no spine near the eyes. It measures up to 6 inches.

THE STICKLEBACKS (*Gasterosteidae*)

Although these fish may be remembered from schooldays as "tiddlers' or "tiddlebats", caught with home-made net and a jam jar in a poc or pond, two of the Sticklebacks also occur either in the sea or brackis water. They are identified by the number of spines on their anterior c front dorsal fin sited before the posterior or hind dorsal fin. The mal builds a nest of plant foliage, seaweeds, by binding the strands togethe with gelatinous threads, produced by his kidneys. A cavity is made i the nest in which the eggs are laid by the female and the male ferociousl guards them until they hatch.

FIFTEEN-SPINED STICKLEBACK (*Spinachia vulgaris*)
Common, especially on the east coast. It frequents rock pools with se weeds, sand pools and shallow waters around our coasts, sometim lurking in water remaining under stones, seaweed, etc., on the sho when the tide has receded. It spawns during May and June. The ma prepares an oval or pear-shaped nest about the size of a man's clenche fist. It is variable in colour, but usually green or olive-green on tl upper-parts, with silvery patterned areas on the sides, a silver line ru ning from in front of and below the eyes with a dark area above it ar on to gill covers; the underside is yellowish. During the breeding seasc the male changes from green to blue, while the female remains gree The rear of the body tapers to a fan-shaped tail fin. An aid to iden fication are the 15 erect spines on the front or anterior dorsal fin. Tl is our largest Stickleback, measuring up to 7 inches.

ΓHREE-SPINED STICKLEBACK (*Gasterosteus aculeatus*)
Jncommon in the sea; common in estuaries, brackish water and pools ıear high water level, also dykes behind seawalls and on salt marshes. Sometimes caught in shrimpers' nets and is also more common in fresh vater. It spawns during May and June. The aggressive male also pre-)ares a similar but smaller nest, into which several females are lured to ay their eggs, after which he guards the eggs and fry when they hatch. It ; also very variable in colour, brownish-grey and green, with bronzy .reas, or silver-white, sometimes pink on the sides. During the breeding eason the male assumes deeper colours and his underside may be orange-ed or scarlet. The three, sometimes four, occasionally only two, separate, rect spines are prominent on the front or anterior dorsal fin. It measures p to $3\frac{1}{2}$ inches.

'HE WRASSES (*Labridae*)
These are some of the most beautifully coloured and patterned fish)und in British coastal waters, particularly in deep pools on rocky, :aweed-strewn coasts. They have a plump, heavily-built appearance, 'ith a single, long dorsal fin along the back, an aid to identification eing the conspicuously thick lips which fold internally into the mouth. he jaws contain rows of strong, sharp teeth used to crush crustaceans ıd molluscs. Wrasse have the curious habit of resting lying on one de on the pool bottom. Some of the Wrasse make seaweed nests in hich to lay the eggs.

ALLAN WRASSE (*Labrus bergylta*) Fig. 28 (a)
ommon. Very variable in its brilliant coloration, but usually of greens, ues, yellows and reds. An aid to identification is that each scale has a hitish or pale bluish centre spot. The vertical fins have yellow or ange rings. It measures up to 10 inches, but larger examples have been ught.

ORKWING WRASSE OR SEA-PARTRIDGE (*Crenilabrus melops*) g. 28 (b)
ery common on south coasts; common elsewhere except in the north, ıere it is uncommon. It frequents shallower pools than the other rasse members and is also often found in colonies. It is extremely riable in coloration, but usually has eight vertical dark brown blotches bands on the red-brown back or around the body, with greenish-low or whitish-green sides becoming paler on the lower sides. The e and snout may have several green, yellow or red marks. The dorsal

(b)

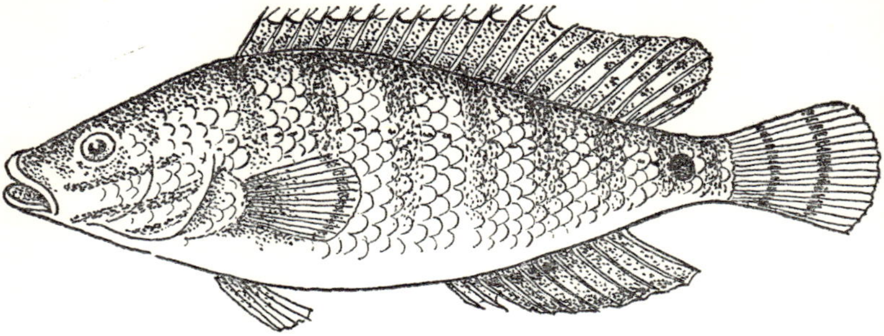

(d)

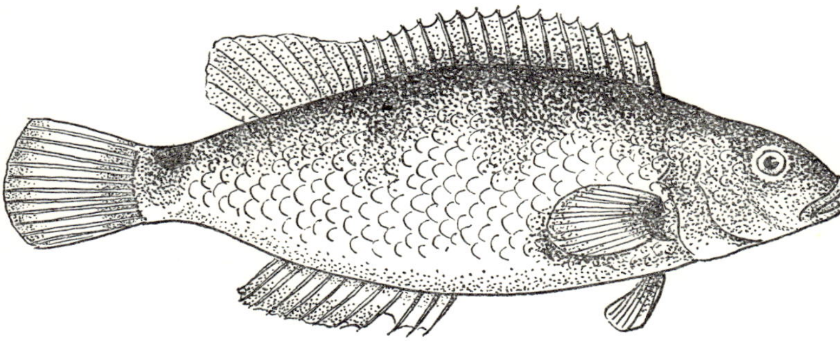

(a)

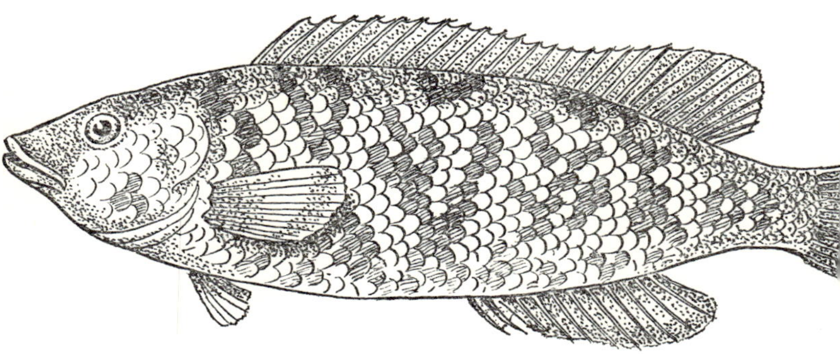

(c)

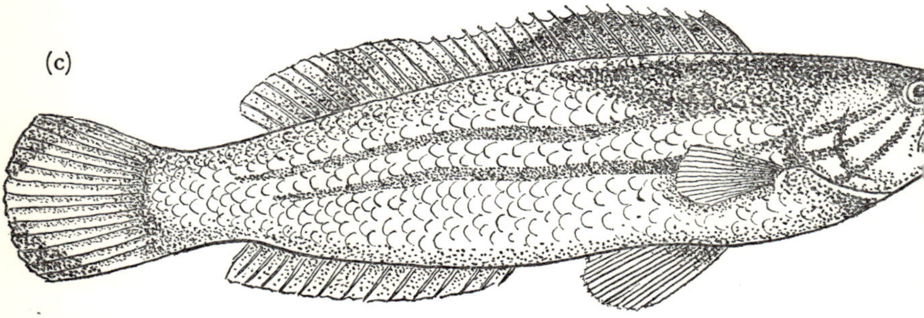

Fig. 28 The Wrasses (b) Corkwing Wrasse *Crenilabrus melops*
(d) Rock Wrasse or Goldsinny *Ctenolabrus rupestris*
(a) Ballan Wrasse *Labrus bergylta*
(c) Cuckoo Wrasse *Labrus mixtus*

fin, anal fin and tail fin may have similar dark brown or red-brown coloration as the body, or pale rings on the tail fin and darker spots on the anal fin. An aid to identification is a single dark spot conspicuously placed on each side of the body close to the tail fin. It measures up to 4 inches.

UCKOO, STRIPED OR RED WRASSE (*Labrus mixtus* Fig. 28 c)

airly common. It is also very variable in its brilliant coloration in both exes. The male is usually yellow or orange, sometimes with a reddish ue or paler on the underside. From each eye several blue bands run ackwards. A dark blue area covers the top of the head and over the houlders including the fore-part of the dorsal fin on the back. The ins are orange or yellow. The female has a reddish tint, the blue bands n the body being absent, but it has several large dark or black blotches n the back beneath and behind the dorsal fin. The fins are also yellow r orange. Like the Corkwing Wrasse it builds a nest of seaweed. It easures up to 10 inches, though larger examples sometimes occur.

OCK WRASSE OR GOLDSINNY (*Ctenolabrus rupestris*) Fig. 28 d)

airly common, particularly on east and south coasts. It is reddish or inkish-gold, darker on the upper-parts. The fins are reddish, the dorsal n having a black blotch on the front portion. There is also a similar ark blotch at the base of the tail. Sometimes it has broad vertical bands r pale patches on the upper-parts. It measures up to 6 inches.

MALL-MOUTHED WRASSE OR ROCK COCK (*Centrolabrus oletus*)

ncommon. It has brown along the upper-parts paling to yellow on the des, and pale silvery-yellow on the underside. The fins are also very-yellow, the front portion of the dorsal fin having a single row dark markings. There is a dull blue area on the gill covers. An aid identification is a dark spot above the eyes. It measures up to 3 inches.

SH EGG CASES

Around our coasts and on the shores the empty egg cases of fish species at normally live in deep water, but spawn in shallower water, are casionally cast up on the shore by the sea. These are also nicknamed Mermaids' Purses". They consist of a horny material, usually flat, ur-sided and roughly oblong in shape, sometimes with coiled threads

at the corners, or stiff projections, used by the female to attach them t
seaweeds or other objects.

Those of the LESSER SPOTTED DOGFISH (*Scyliorhinus cani-culus*) are about 2 inches in length, excluding the coiled threads, whil those of the GREATER SPOTTED DOGFISH (*Scyliorhinus stellaris* Fig. 29 (a) measure up to 4 inches, excluding the coils. Both are dar coloured and have long coiled threads or tendrils from the four corner of the egg case, although examples are sometimes found in which thes are missing.

(a)

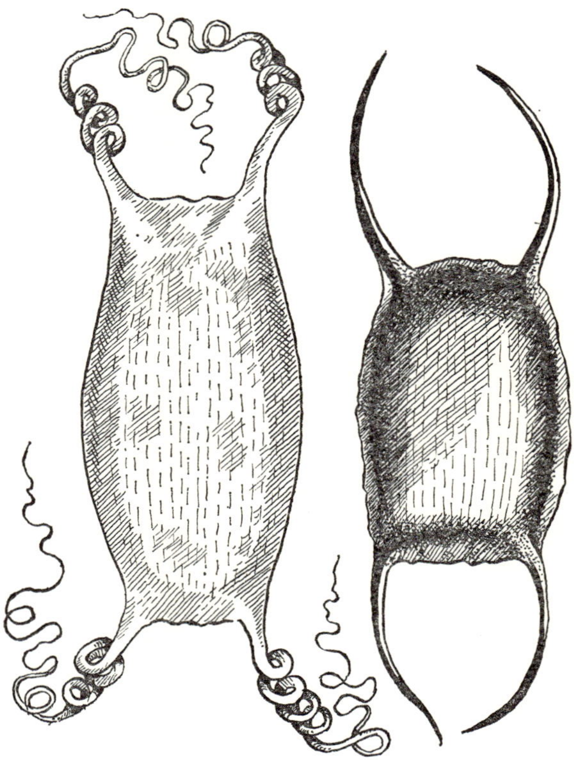

(b)

Fig. 29 Fish egg cases (a) Greater Spotted Dogfish *Scyliorhinus stellaris* (b) Common Skate *Raja batis*

The empty, horny egg cases of the Rays and Skates are roughly re angular with the four corners forming stiff, horn-like projections points which are hollow tubes, but do not have coiled threads like t Dogfish. The four horns or fine points are usually two long and t short, one of each at each end of the case. The CUCKOO RAY (*Raja naevus*) produces egg cases about 2 inches in length, excluding the points, one pair of these being considerably longer than the other pair twist across each other. The LOCK-NOSED RAY (*Raja oxyrhynchus*) produces cases approximately 5 inches in length excluding the points which are about equal in length. The SPOTTED RAY (*Raja montagui*) produces cases about 3 inches in length, excluding the points, one pair being approximately twice the length of the other pair, each point having a slit along its outer edge. The THORNBACK RAY (*Raja clavata*) produces egg cases 3 inches in length, excluding the points, all of which are about equal in length. It is convex on one side and sometimes has a mat of short hairs if only recently cast ashore. The COMMON SKATE (*Raja batis*) Fig. 29 (b) produces cases up to 6 inches in length, excluding the points, which are about equal in length but the pairs turn inwards. It sometimes also has bunches of sticky hairs on each side of the egg case.

Chapter 5

Crustacea – Crabs, Spider Crabs and Pycnogonida – Sea Spiders

The Crabs are large crustacea and members of the group *Decapoda* as described fully in Chapter 6. Each crab begins its life cycle as an egg The female carries her egg mass about with her, totalling up to 20,00(very small eggs in some species, on the underside of her body, affordin; them some protection from predators until the larvae hatch.

These are extremely tiny, floating about in the upper levels of th sea and not moving to the sea bottom until they are nearing or in th adult stage. The microscopic larvae "swim" or jerk themselves througl the water, feeding upon decaying matter. They themselves are a sourc of food to fish and other marine creatures.

Several stages are passed through. Each species varies, but basicall they have a large, bulbous head, prominent eyes, a curved "horn" o the front parts and a pointed flexible tail and plumed legs. As it grow it casts its skin until each larva finally becomes a pupa or magalop from which it emerges as a minute but recognisable crab.

The crab's shell or carapace becomes too small for it as its body grow inside it and being non-expandable this has to be cast off periodicall When this takes place the crab is most vulnerable to attack, its soft par having no protection while its larger limey shell or carapace is bein produced and hardening, so the crab must find a hiding-place for th transformation among rocks and behind seaweed and lurk there unt it is able to fend for itself again.

The crab's body is nearly round, oval or triangular in shape. It h four pairs of legs which are used for swimming and walking, while the front part of the body are the two large, hard claw limbs arme with the nippers, the latter being used for fighting, defence, and gathe ing food.

Crabs frequently fight among themselves and during the battle o or more limbs may snap off, or a crab may become trapped by a lim

Then the crab will deliberately bend the limb in such a way that it breaks and the crab can thus free itself to escape.

The comic seaside postcard depicting the thin man or fat lady hopping about painfully with a large crab, gripping with a nipper, dangling from one of their toes is not entirely fiction. The nippers in some crab species are particularly powerful and able to inflict an injury.

Gulls and the wading sea birds suffer from the attentions of the larger species of crabs which attack these birds' feet and legs. It is not unknown for a crab to amputate a birds' foot or leg, and this probably accounts for most cases of one-legged gulls seen on the coast from time to time.

Other features of the crab's anatomy are the eyes fixed on stalks with all-round vision, "ears" on its feelers which enable it to hear better through water than it can when in open air, and glands on the feelers which act as the crab's nose so that it can detect living or dead creatures.

Two species of crab are hosts to parasites. On the under-tail region of the Shore Crab (*Carcinus maenas*), both male and female, a light brown coloured mass discloses the presence of the common parasitic Barnacle, SACCULINA CARCINI. It occurs in other Crab species, but only rarely. The under-tail region of the Hermit Crab (*Eupagurus bernhardus*) may have a yellow, sometimes orange, lengthy lump which shows the presence of the fairly common parasitic Barnacle PELTOGASTER PAGURI. These parasites also prey on other Decapod Crustaceans. They can be identified from the crab's egg-mass, as the parasite's mass is non-granular.

BROAD-CLAWED PORCELAIN CRAB (*Porcellana platycheles*) Fig. 30 (a)

Common. It frequents the middle and lower shore, particularly muddy, gravelly, stony shores, where it lurks under stones. In colour it is greyish-brown or a red-brown, though its dull appearance often matches its habitat. The broad, flat nippers appear huge in comparison with the small body and length of the other legs. It has four pairs of legs, similar to other crabs, but three pairs may only be apparent, the "missing" pair being smaller and folded across the base of its "tail". The reason for this behaviour is unknown, but may be a form of "rear end" protection. The edge of the shell is hairy, so, too, are the legs and nippers. Particles of mud collect on the hairs on its nippers and legs and assist

(a) (b)

Fig. 30 (a) Broad-clawed Porcelain Crab *Porcellana platycheles*
(b) Long-clawed Porcelain Crab *Porcellana longicornis*

in camouflaging the crab when it shelters under a stone. The shell or carapace measures up to 1 inch in width and slightly more in length.

EDIBLE CRAB (*Cancer pagurus*)
Common. It frequents the middle and lower shore, especially rocky coasts where there is an abundance of stones and seaweeds to hide under. The coloration is brown-pink the nippers being darker. The latter are strong and powerful and can give a sharp pinch to discourage handling. Pugnacious in temperament, if placed on an open area it will usually not attempt to escape, but will "stand and fight". As the adversary moves around the Edible Crab it will raise its nipper legs hold them in a fighting attitude and also turn its body so that it is always facing the foe. Any attempt to sieze it is met with immediate retaliation. This is the crab you see on sale on fishmongers' slabs. It is an important species commercially. The soft, yellow liver, called the "fat", is considered a delicacy. The shell or carapace is often adorned with Acorn Barnacles and tubes of Sea Worms. It measures up to 10

inches across the hard, almost oval-shaped shell or carapace. Large specimens, however, are not usually found on the shore unless accidentally cast there, being inhabitants of deep water. Examples of 4 to 5 inches are more likely in rock pools.

FIDDLER OR VELVET SWIMMING CRAB (*Portunus puber*) Fig. 31

Common, more so on south and west coasts of England. It frequents the lower shore in rock pools under stones. The coloration is usually, reddish-brown, occasionally dark brown, the joints and legs being striped or marked with blue, the nippers partly blue and partly the colour of the shell. Pugnacious in attitude when disturbed despite small size. It should not be placed in a marine aquarium as it will destroy other inhabitants. When confined together it is cannibalistic and will kill and devour its own kind. This crab can swim through the water, unlike other crab species which crawl on the mud or sand shore. The last joint of the hind legs is flattened into rounded-oval "plates", these having a fringe of hairs on them, which are used by the crab to swim. The movements it makes while doing so have been likened to the playing of a violin, hence its name Fiddler Crab. It has red eyes. The other name, Velvet, is derived from the covering of short, soft, downy hairs on its shell. It measures up to 4 inches in width across the shell.

Fig. 31 Fiddler or Velvet Swimming Crab *Portunus puber*

FURROWED CRAB (*Xantho incisus*) Fig. 32

Locally common, chiefly on the south-west coast of England. It frequents the lower shore and rock pools, beneath stones and debris. In colour it is brown with a reddish tinge, the nippers being black. The uneven shell is up to 3 inches in width. It derives its name from a pronounced groove between the eyes in the middle of the shell.

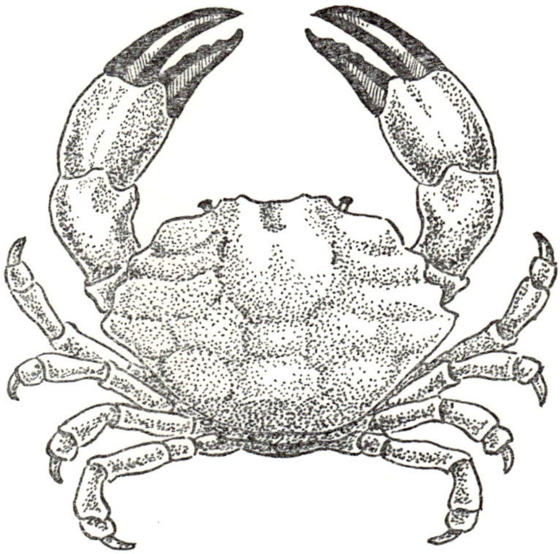

Fig. 32 Furrowed Crab *Xantho incisus*

HAIRY CRAB (*Pilumnus hirtellus*)
Common, chiefly south and west coasts of England. It frequents the lower shore, lurking under stones and seaweeds. In coloration it is brownish-red, the nippers being paler brown. An aid to identification is that one nipper is noticeably larger than the other, also the crab is entirely hairy, hence its name. It measures up to an inch in width across the shell.

HERMIT CRAB (*Eupagurus bernhardus*) Fig. 33
Very common. It frequents the lower shore in pools on sandy flats and stony or rocky shores, particularly where there are numbers of vacant Whelk and other Univalve shells. Its coloration is reddish and yellow with a brown abdomen. It usually measures up to 3 inches in *length,* occasionally more. Easily identified because of its body shape and behaviour which is worth a detailed description. Only the front portion of the shell or carapace and legs are protected with hard covering; the asymmetrical abdomen and tail are soft and unprotected and not tucked under the body. To protect its "rear end" from attack by enemies the

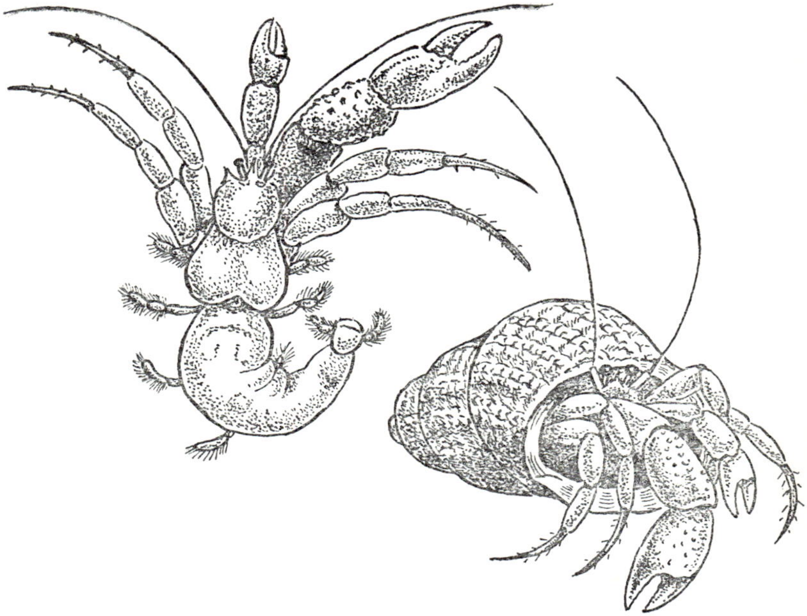

Fig. 33 Hermit Crab *Eupagurus bernhardus*

Hermit Crab must find a suitable vacant whelk or other shell. When it does so it reverses into the shell, the tail fitting into the shell's spiral curves. The Hermit Crab hauls and carries this shell "home" along behind it wherever it goes, the legs being more powerfully developed to do so. When it needs to, the Hermit Crab can withdraw itself almost entirely inside its "home" leaving only a short length of its right nipper protruding to close the entrance. Eventually the young Hermit Crab has to vacate its protecting shell, probably a Top Shell, Periwinkle, Dog Whelk, etc., as the tail grows too big for this. When the crab finds another, bigger shell – usually a Common Whelk shell is appropriated by large Hermit Crabs – it drags itself close to the new intended "home", inspects it to discover if there is already another tenant and if not, it removes its soft body from the old shell and gets into the new one very rapidly. If it is too slow about this task a predatory fish would lose no time in seizing the unprotected portion of the crab. The Hermit Crab may change its shell "home" a dozen times or more, in its lifetime, gradually increasing them in size. This crab is extremely pugnacious and will use its powerful nippers to protect itself when attacked from a frontal position.

The Hermit Crab frequently forms a "partnership" with a Sea Anemone, particularly ADAMSIA PALLIATA and CALLIACTIS PARASITICA. The anemone attaches itself to the crab's whelk shell "home". This is to their mutual benefit. The crab obtains extra protection because fish and other enemies will not attempt to attack the Hermit Crab and come within range of the anemone's poison tentacles. The anemone gets some of the crab's food. When the crab kills a victim or tears apart the body of an already-dead marine creature, some of the particles float upwards and are siezed by the anemone, the latter apparently not caring that it is carried along continually on the sea bed or pool bottom while the crab is foraging. Alternatively, the anemone may sting and kill a victim which falls close to the crab below. The latter proceeds to use its powerful nippers to cut up the corpse into portions, some of which is passed upwards to the waiting outstretched tentacles of the anemone.

It is almost impossible to extract an unwilling Hermit Crab from its shell, without tearing it in half. One way is to break off the tip of the univalve shell and prod the soft tail of the crab gently. It may think an enemy has found a way in and exits the shell voluntarily. Less kindly is to place a lit match under the shell whereupon the crab usually vacates its "home" or releases its grip, and can be extracted. On the pool bottom another way is to place the crab and shell close to an anemone. Despite the fact it will live in harmonious partnership with an anemone on its own "shell home" another anemone affixed to a rock or object has no such feelings of friendly cohabitation and will eat the crab if it gets the chance. Realising this, the Hermit Crab vacates its whelk shell as quickly as possible leaving the shell in the anemone's grasp, while the crab hurriedly seeks any form of shelter for its tail.

The Hermit Crab's shell is also "colonised" by Sponges, Sea Firs (Hydroids) and Acorn Barnacles. Even more curious, the Hermit Crab sometimes allows a Marine Worm (*Nereis fucata*) to share what little space there is inside its shell "home".

LONG-CLAWED PORCELAIN CRAB (*Porcellana longicornis*) Fig. 30 (b)

Common. It frequents the lower shore in stony habitats and pools, lurking under stones and large seaweeds. In colour it is brownish-red, the shell or carapace having a shiny appearance. The nippers appear huge in comparison with the body size and leg length. It has a pair of thin antennae on the head about twice the length of the shell. Like the Broad-Clawed Porcelain Crab it has four pairs of legs, but only three pairs may be apparent, the "missing" fourth pair being small and folded across the base of its tail. It is also hairless, which is an aid to identify it from its relative. Less likely to be found in muddy habitats where the Broad-Clawed Porcelain Crab prefers to hide. Also smaller than its relative, the nearly round shell being up to ½ inch in width.

MASKED CRAB (*Corystes cassivelaunus*) Fig. 34
Fairly common on south coasts, rarer elsewhere. It frequents the lower shore on sandy flats and estuaries. It is brownish-yellow or a dark straw colour. The name is derived from the markings on the shell which, with the shape, give it a fancied resemblance to the features of a human face or a mask. The claw limbs are long and narrow, with small nippers, the two hairy antennae or "feelers" being very long and extending beyond the claw limbs and nippers. It is most likely to be discovered while you are digging, perhaps for bait, due to the crab's habit, as the tide recedes, of burying itself in the sand. Here it remains until the tide

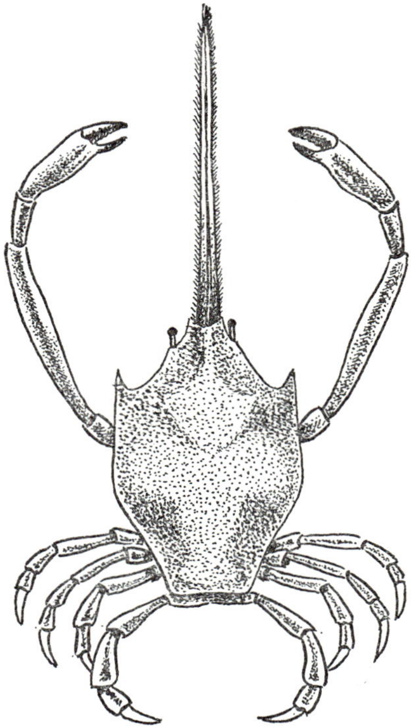

Fig. 34 Masked Crab *Corystes cassivelaunus*

turns, with the two feelers or antennae held together, to create a tube, the tips of which, with its eyes, project above the sand surface. A small crab, the male being an inch in width and $1\frac{1}{2}$ inches in length, the female being about half these measurements. Both are quickly recognisable because the shell or carapace is longer than it is wide, the opposite of the typical crab-shape, which is wider than it is long.

PEA CRAB (*Pinnotheres pisum*) Fig. 35
Fairly common on south and west coasts, rarer elsewhere. It frequents shores, estuaries and coasts where there are also Bivalve Molluscs, Mussels, Horse Mussels, Cockles, etc. The male is creamy-yellow, pale buff-brown or yellowish-grey, the female is semi-transparent brownish, with a yellow tail and yellow area on the front of the shell. It has the curious habit of residing inside the shells of a living Mussel, Cockle, Pinna or other shellfish as a lodger, even while the rightful tenant is still in occupation. Apparently the shellfish does not mind this invasion, as the crab does it no physical harm. Possibly the partnership has a two-fold benefit, like the Hermit Crab and Sea Anemone, though it is not a case of the armed Pea Crab defending the defenceless shellfish against a predatory intruder, when the shellfish opens its shell to feed. If a mussel or cockle is opened and the Pea Crab revealed it will usually either tuck all its legs close around its shell and lie still pretending to be dead, or it may leap out of its lodgings and hope to escape. Possibly it acts as a scavenger inside its host's shell. It is known to take the filtered

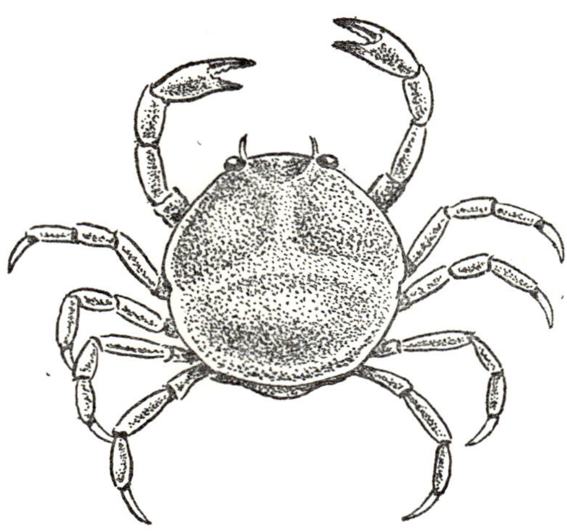

Fig. 35 Pea Crab *Pinnotheres pisum*

food materials off the shellfish's gills, thus stealing the food from its host. A very small species, the shell or carapace being almost round in shape, both details giving rise to its name. The male's shell is up to $\frac{1}{4}$ inch in width, the female's shell is about $\frac{1}{3}$ inch in width.

SHORE OR GREEN CRAB (*Carcinus maenas*) Fig. 36
Very common and the commonest crab around Britain. It frequents the middle and lower shore and occurs on mudflats, in creeks, in estuaries, on sandy, gravelly, chalky, rocky and other types of shore and coast, particularly under seaweed covered stones and rocks. It is very active and frequently met with, away from hiding-places, at the tide's edge, feeding on other living marine creatures, Sandhoppers, Bivalve Molluscs, etc., or carrion. The Shore Crab is able to leave the sea or pool for considerable lengths of time, being able to breathe on land if its gills remain moist. This crab is often found dead in hundreds along the shore, cast there among seaweed fragments and other sea debris. It is a nuisance to sea anglers and those fishing from piers, docks, in salt-marsh dykes, etc., because it steals the bait and a movement on the line proves to be, not a fish, but a Shore Crab clinging to the baited hook. The adults are usually a darkish-green, hence its second name. Smaller, young Shore Crabs vary from muddy brown, greenish-brown to a

Fig. 36 Shore Crab *Carcinus maenas*

yellowish-buff. It is edible and in the past has been gathered for food, but as it is smaller than the Edible Crab and therefore less fleshy it is seldom so-used today. The adult shell measures up to 4 inches in width.

SPIDER CRABS

They were given this name because of a similarity to a spider, more apparent in some species than in others, but apart from this they have no physical relationship with the land spiders. Excluding the Spiny Spider Crab the body is small; to it are joined long, jointed legs and claw-limbs with nippers. The Spider Crabs have an approximately triangular-shaped body, shell or carapace, with the head forming the apex. The legs are large in comparison with the body size, also being very spindly in some species. The Common Spider Crab and Spiny Spider Crab have spines or protuberances on the back of the shell to which they affix small pieces of seaweed, sponge and similar. The reason for this is camouflage, to disguise themselves when living among such seaweeds and sponge. If the Crabs remain motionless when clothed in such a fashion they are difficult to see, by the human and their predator enemy's eye. They will even change their disguise to suit their background. If they crawl among, or are placed in, brown seaweeds while "wearing" green seaweeds, for example, they will quickly remove all the green seaweeds that now make them conspicuous and replace them with brown seaweeds to blend again with their surroundings. Spider Crabs occur in the same habitats as crabs, but also occur in deep water. Sometimes they are trawled up with shrimps in shrimpers' nets.

COMMON SPIDER CRAB OR SEA TOAD (*Hyas araneus*) Fig. 37

Fairly common. It frequents the lower shore where it occurs under seaweeds and in pools. In coloration a dull reddish-purple. It is spineless but has some short, hook-like protuberances or tubercles projecting on its back to which it fixes small pieces of seaweed, such as Coralline, and Sponge and Sea Firs, the purpose being to camouflage itself, which it does successfully. The thin legs are longer than the thicker nipper limbs. The long triangular shell or carapace is up to 2 inches wide at its hind end.

LONG-LEGGED SPIDER CRAB (*Macropodia rostrata*)

Common. It frequents the lower shore, under seaweeds, stones and rocks. In coloration reddish or brownish-yellow. The hairy legs are considerably longer and thinner than the nipper limbs. The second pair are

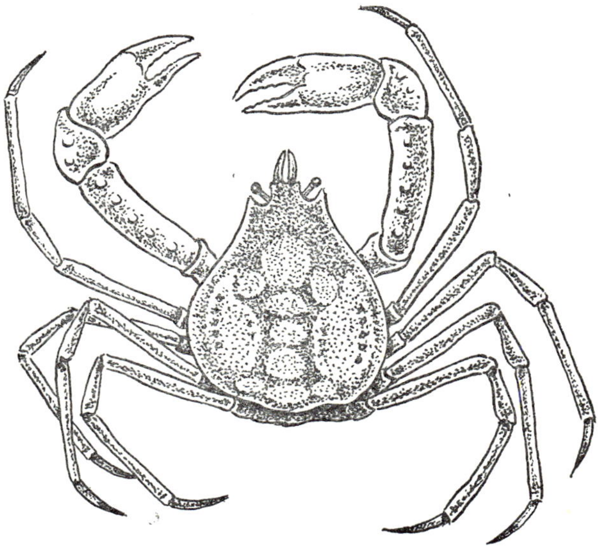

Fig. 37 Common Spider Crab *Hyas araneus*

ıbout three times the length of the shell, all of these giving the crab a ɋpider-like appearance. The shell has several spines and measures up ɔ ½ inch wide at the hind end.

ꞏCORPION SPIDER CRAB (*Inachus dorsettensis*) Fig. 38
'airly common. It frequents the lower shore, in pools and under sea-ᵥeeds. It is brownish-yellow in colour. The triangular-shaped shell with ɔunded corners bears several small warts. The legs are longer and ıinner than the nipper limbs, the second pair being very long. It ·equently covers itself with pieces of seaweed, sponge, etc. Despite name more spider-like than scorpion-like. The shell measures up to ¾ inch ·ide at the hind end.

PINY SPIDER CRAB, THORNBACK CRAB OR DEVIL'S CRAB
Maia squinado) Fig. 39
ommon, more so south and west coasts. It frequents the lower shore, pecially sandy coasts, under rocks, stones and seaweed. In coloration is reddish and yellowish-brown. The hairy legs are long and narrow, hile the shell is covered with numerous sharp, hard spines and short ɔoks, projecting in different directions. Upon these this crab fixes tufts seaweed, sponge, etc., to hide itself from predators. It is edible and

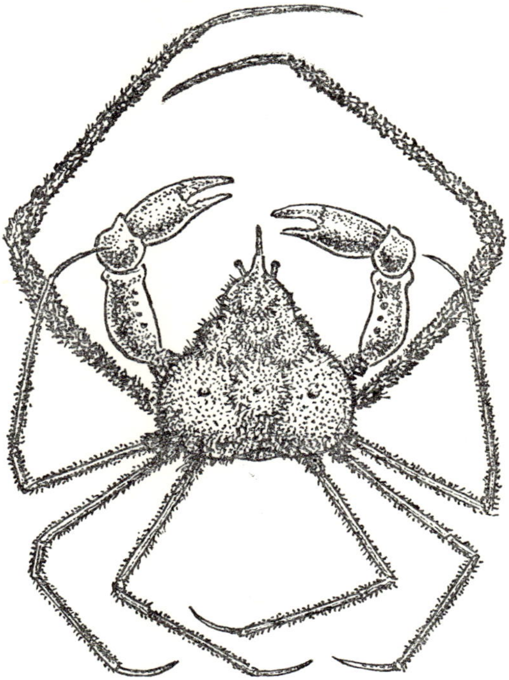

Fig. 38 Scorpion Spider Crab *Inachus dorsettensis*

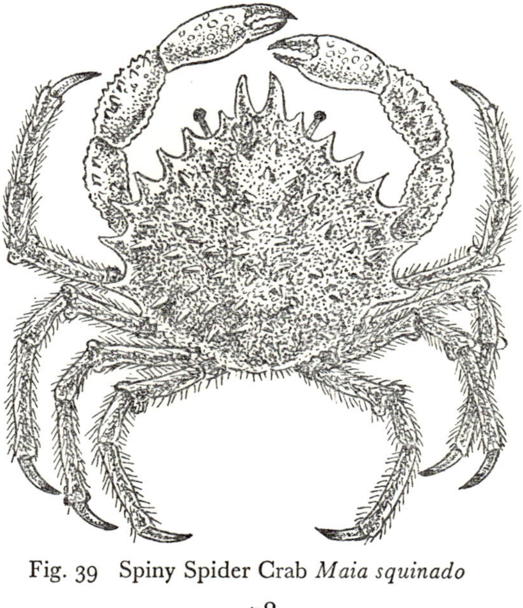

Fig. 39 Spiny Spider Crab *Maia squinado*

has been gathered for food in the past, sometimes being trapped accidentally in lobster pots, but the flesh is coarse and of poorer quality than the Edible Crab. This is the largest British Spider Crab, the shell measuring up to 7 inches in width and 9 inches in length.

SEA SPIDERS (*Pycnogonida*)

The Sea Spiders are members of the group *Arthropoda* and have four pairs of legs, but are not particularly similar to the Land Spiders. The Sea Spiders have very small, thin bodies, the abdomen, is minute and rudimentary, some branches of the stomach penetrating into the tops of the Sea Spiders' legs. On the head there are two pairs of appendages used for feeding. They are carnivorous and mainly prey on Sea Anemones and Sea Firs.

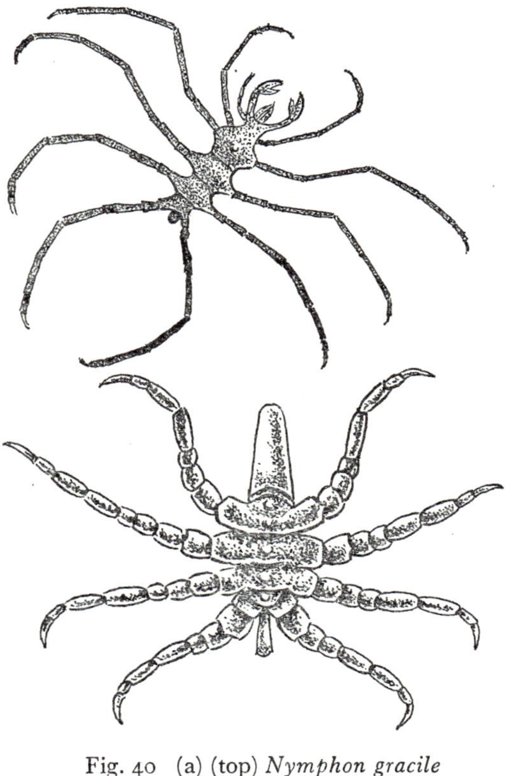

Fig. 40 (a) (top) *Nymphon gracile*
(b) (bottom) *Pycnogonum littorale*

ENDEIS SPINOSA

Fairly common. It frequents the lower shore, in pools, on Sea Firs, Sea Anemones and Seaweeds. In coloration it is either red or greenish-brown. Due to its size it is extremely difficult to detect. Its body length is up to $\frac{1}{5}$ of an inch, the longer legs measuring up to $\frac{4}{5}$ inch in length.

NYMPHON GRACILE Fig. 40 (a)

Common. It frequents the middle and lower shore, in rocky pools, under stones and seaweed. In coloration red, pinkish-yellow or pink. The body is very slim, measuring up to $\frac{1}{3}$ of an inch in length, the legs being up to 1 inch in length.

PYCNOGONUM LITTORALE Fig. 40 (b)

Fairly common. It frequents the lower shore, in pools inhabited by Sea Anemones on which it may be seen crawling dangerously or feeding; also lurks under stones, seaweed and sea debris, but is sluggish by day due to a sensitivity to light. In colour it is light brown or brownish-yellow. The thick, segmented legs, which give them a swollen, "knobby" appearance, have hook-like claws at their tips. The body measures up to $\frac{3}{5}$ inch in length.

Chapter 6

Crustacea – Various

The *Crustacea* are a fairly large group of varied marine or semi-marine animals, which belong to the animal group *Arthropoda,* from the Greek for "jointed-legged animals". They have a head and thorax, which may be divided, but is sometimes indistinguishable if there is no dividing "neck". But usually the long abdomen is distinct, more recognisable and thought of as the region of the tail. The limbs vary in shape and size and are used for swimming, crawling, burrowing, fighting, surveying and securing their food and sometimes breaking it into smaller portions before consumption. Several species are also armed with nipper limbs for protection or obtaining prey. Sometimes they may become accidentally trapped or an enemy may seize a limb, but the *Crustacea* can muscularly bend the limb to such a position that it breaks off at a special weak point and they are released. A new limb will eventually grow.

The soft body has a hard, outer covering and from this the *Crustacea* get their name, the word being derived from the Latin "crusta", a crust, referring to the protective, horny, crust-like shell or carapace. Another way to describe it is as an external skeleton. It is necessary for some species, as their body grows too large within it, to shed their hard unstretching carapace, as described in the Crab chapter. Some of them, such as the adult Barnacles, large Lobsters, minute Sandhoppers, are so unlike as to be thought unrelated, but at some stage in their early life they have had, or in the adult stage still have, an anatomical similarity to connect them together as all members of the *Crustacea.*

They begin life as an egg and may have a free-swimming larval stage or be retained in a brood pouch by the female until sufficiently developed. The Crustaceans are nature's refuse collectors of the sea and shore. They are scavengers, with a stomach which has strong muscles that operate "teeth" to pulp and digest the food, after it has been shredded by the limbs. They consume vast quantities of dead and decaying animal and vegetable matter which would otherwise foul these habitats.

BARNACLES

These animals, the *Cirripedia,* are common on rocky coasts and shores The hard, limy plates or shells of certain species, the Acorn Barnacles, form dense, wide, encrusting areas. At first glance they could be mistaken for miniature Limpets, but close examination will show they have several ribbed plates, enclosing the body, and are attached to the rock or other surface by their base. At low tide, when exposed, the animal closes its plates to retain sufficient water and air until again covered by sea water. When it is submerged the plates open and six pairs of curved, feathery legs, like the bent fingers of a hand, are continually extended and withdrawn in a grasping motion. This action creates a current of water which traps the minute organisms on which it feeds and carries them into its mouth.

The Acorn Barnacle begins life as an egg, from which a tiny larvae emerges that lives amongst the plankton. Here it swims about, most unlike the adult, with three pairs of legs, head, eyes, a thin shield and tail. It moults several times during its life cycle and finally, after discovering a suitable surface – rocks, stones, breakwaters, shells, other crustaceans – attaches itself to it with its head downward, with a kind of cement. Here the head and tail waste away. But the mouth and other vital organs remain and the necessary protective plates and food-collecting legs develop. The plates of some species are so sharp that they can badly cut the feet of anyone who tries to walk barefooted over them. It is extremely difficult to distinguish between some species of Acorn Barnacle, one method being to identify the shape of the aperture at the top where it is surrounded by the plates.

The Ship, Stalked or Goose Barnacle has a similar life cycle, but they usually attach themselves to ships' hulls – where in large numbers they create a 'slowing' nuisance and have to be regularly scraped off – or to wreckage, floating timber, buoys, etc.

ACORN BARNACLES

BALANUS BALANOIDES Fig. 41 (a)

Common, more so on northern coasts than in the south. It is found between the higher upper shore and lower shore, where it occurs in large numbers on exposed rocks, stones, chalk boulders, harbour installations, etc. It has six plates which are so formed as to make a

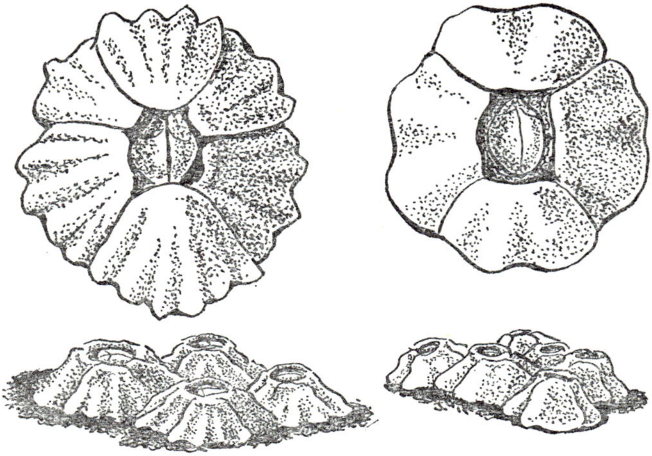

Fig. 41 Acorn Barnacles (a) (left) *Balanus balanoides*
(b) (right) *Elminius modestus*

approximate "diamond" shaped aperture when viewed from above. Plates are dirty white in colour. It measures $\frac{1}{2}$ inch in height and up to $\frac{1}{3}$ inch in width, but size varies considerably depending on habitat.

BALANUS CRENATUS

Common. It frequents the lower shore, where it occurs on the overhanging ledges of rocks, sides of stones, breakwaters, etc., rather than on the exposed surfaces. The six ridged plates lean towards one side and are light grey in colour. May be identified, by the way a white ring is left on the rock surface after the barnacle is dislodged. It measures up to $\frac{1}{2}$ inch in diameter.

BALANUS PERFORATUS

Common, but mainly on south and west coasts of England. It frequents the lower shore where it occurs on the overhangs of rocks, and in the crevices. The six plates are almost smooth or lined vertically and are purple-brown in colour although sometimes they may be very light-coloured with only a tint of purple. The adults are easily identified by their size, this being the largest British Acorn Barnacle, measuring up to $1\frac{1}{4}$ inches in height and $\frac{3}{4}$ inch in width.

CHTHAMALUS STELLATUS

Common, but occurs only on south-west and west coasts of England, elsewhere it is rare or absent. It frequents the upper shore and middle shore, on exposed rocks, stones, breakwaters, etc. The six plates are so formed as to make a "kite" or "shield" shaped aperture when viewed from above. Plates are white or light grey. It measures up to $\frac{1}{2}$ inch in height and up to $\frac{1}{3}$ inch in width.

ELMINIUS MODESTUS Fig. 41 (b)

Very common on south, south-west and parts of the east coast of England but yearly increasing its range northwards. Particularly found in muddy estuaries. This barnacle originated from New Zealand and Australian estuaries and is believed to have been introduced into Britain probably attached to ships' hulls about 25 years ago. It frequents the middle shore but not in such exposed positions as other species. The four plates are a semi-transparent, pearl grey colour.

VERRUCA STROEMIA

Fairly common. It frequents the lower shore where it occurs on shells, under stones and other objects. It does not colonise a large area as a group of barnacles, but is usually found alone or as scattered individuals. An aid to identification is that the four longitudinally ridged plates are all different in size and are white in colour. It measures up to $\frac{1}{4}$ inch in width.

STALKED BARNACLES

GOOSE OR SHIP BARNACLE (*Lepas anatifera*)

Common. It is one of the Stalked Barnacles. It attaches itself, sometimes in large numbers, to ship's hulls, wreckage, buoys and other floating objects in the sea, and may be washed ashore attached to them. It can also be found on rocks, pier and jetty timber uprights, breakwaters, etc. Goose Barnacles have a similar plate construction, but these are not so firm and hard as in Acorn Barnacles. In the final larval stage it develops a mantle and almost-triangular shaped shell of several plates, its body being held together by a membrane. The long, flexible, fleshy stalk is provided with powerful muscles, by which it attaches itself to hang down, in clusters, from the habitat. The glands secrete an adhesive liquid used to anchor itself firmly and almost immovably in this way. The long stalk is blue-grey, the rather triangular limy plates being whitish or pale bluish. It gets its name from a curious belief in the past that the Goose Barnacle was the first stage in the development of the

Barnacle Goose and that the "Goose" grew attached to the shell of the barnacle. The belief may have arisen because of a fancied resemblance of the barnacle's stalk to that of the sea bird's legs and the barnacle's body and feathery tentacles inside or projecting from the shell to the feathers of the Goose. The stalk and plates combined measure up to 4 inches in length.

LEPAS FASCICULARIS

Infrequent, usually on south and south-west coasts, after being washed ashore. This Stalked Barnacle is more likely to be found alive drifting in the sea. It creates a white, spongy float to attach itself to from a mucus secretion in which air bubbles are trapped, produced by the short, blue-grey stalk, the latter being variable in length. The thin, frail, semi-transparent bluish-white plates, excluding stalk, are about $1\frac{1}{2}$ inches in length.

The Prawns, Shrimps, Lobsters and Crabs are members of a group of the larger *Crustacea* known as the *Decapoda* and there is really no need for me to add how relished these are for human consumption. The name comes from the Greek for "ten legs", as they have five pairs of jointed "walking legs" borne by the segments of the thorax, between the head and the tail, the soft body being protected by the harder carapace. There may also be a pseudo-pair of legs on the head, used for feeding. The first pair of true legs are sometimes armed with small or large nippers, while the hinder segments may have several pairs of appendages or swimmerets, the last, the tail segment, forming a fan-shape. The *Decapoda* is further divided up into two groups – the *Natantia* or Swimming Decapods – the Prawns, and Shrimps; and the *Reptantia* or Crawling Decapods, the Lobsters and Crabs; the separation being self-explanatory in the titles.

The Prawns come into the shallower water during the summer, where they may be found in rock pools, but retreat to deeper water during the winter months. Their eyes are mounted on stalks which can be moved around with the two pairs of antennae or feelers, to sweep the area and detect items of food. The hard protective carapace from the head extends forward between the eyes to form a beak-like projection called the rostrum. Even when seen in a pool prawns are not easy to catch and as soon as the net is placed near them they "back-pedal" away from it, by bending the abdomen and extended tail fan under the body so that they dart rapidly backwards. In early summer the female produces the eggs which, after being fertilised by the male, are carried by her attached to hairs on some of the swimmerets, hatch into tiny larvae that join the plankton for a short time.

Although they appear to resemble Prawns at first glance it will be seen that Shrimps do not have the beak-like projecting rostrum, also they have only one pair of antennae, not two. They frequent the sandy and muddy shores, estuaries and mudflats, into which they burrow, more so than the rock pools. While buried a portion of their antennae or feelers remain out of the sand, so that the Shrimp can continue to seek food, being carnivorous on other crustaceans, marine worms, molluscs, fish eggs, as well as being scavengers. Shrimps are more likely to be buried in the sand by day, being nocturnal, reappearing to walk about and scavenge in the darkness. They do not swim so regularly as Prawns. Shrimps reproduce by eggs in a similar manner to Prawns.

The Lobsters are also similar in anatomical shape to the Prawns and Shrimps, but are mostly considerably larger, armed with formidable nipper limbs, the stronger body being protected by a tougher, thicker, limy "shell" or carapace. Common Lobsters crawl more than Prawns and Shrimps, but by quickly bending their abdomen and tail under their body and flexing them straight the thrust causes the Lobster to retreat backwards and this action can be maintained for quite long distances. The female Lobster carries the fertilised eggs, attached to the underside of her abdomen or "tail", for a year before the free-swimming larvae emerge from them and after a short period descend to the sea bottom. Lobsters are scavengers and also carnivorous on other marine creatures. They are lured into the familiar lobster pots with dead fish and other baits.

PRAWNS

AESOP PRAWN (*Pandalus montagui*) Fig. 42 (a)

Common. It frequents the lower shore, near the sea, where it occurs in pools. The outer antennae is much longer than the body and has pale and dark red rings. The body is reddish-grey, with red markings and dots and has a transparent aspect. It measures up to 2½ inches in length.

CHAMELEON PRAWN (*Hippolyte varians*) Fig. 42 (b)

Common, except in Scotland. It frequents the lower shore where it occurs in pools among the seaweeds. The outer antennae is approximately half the length of the body, the latter, in the middle, having the appearance of a humped back. Its name is derived from its varia-

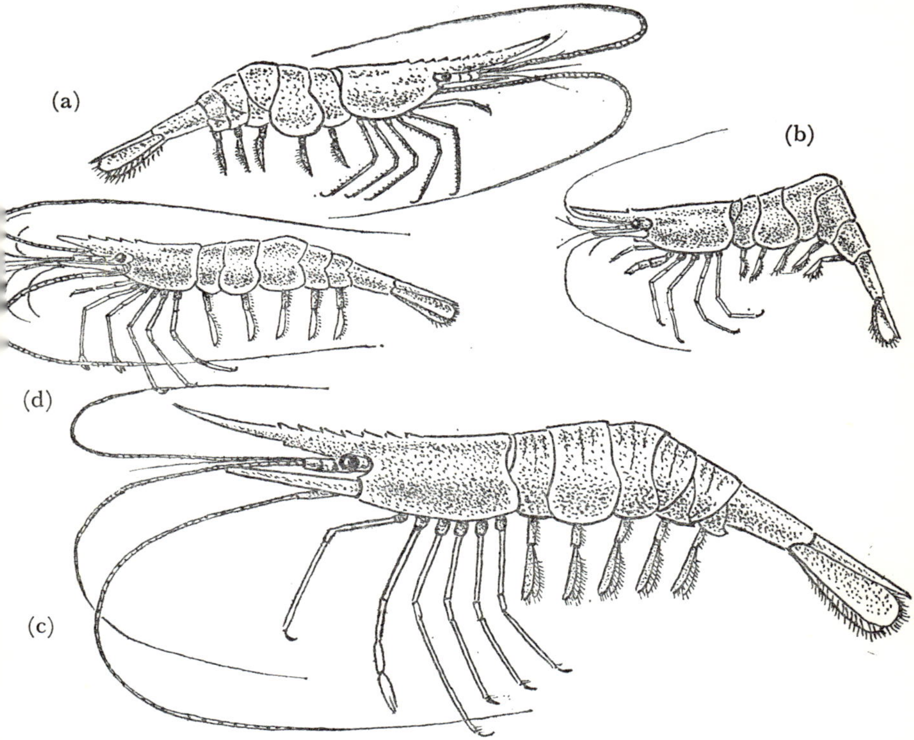

Fig. 42 Prawns (a) Aesop Prawn *Pandalus montagui*
(d) *Palaemonetes varians* (b) Chameleon Prawn *Hippolyte varians*
(c) Common Prawn *Leander serratus*

tions of colour. During the daytime it is usually a greenish colour, sometimes reddish or brownish, but at night it is a transparent bluish colour. These protective colour changes depend on the type of seaweed it is living among and needs to blend with. If it is placed among red or brown seaweeds a green Chameleon Prawn will gradually change to red or brown and vice-versa. It measures up to $\frac{3}{4}$ inch in length.

COMMON OR EDIBLE PRAWN (*Leander serratus*) Fig. 42 (c)
Common on south and west coasts. It frequents the lower shore where it occurs in larger pools near the sea among the seaweeds. The outer antennae is $1\frac{1}{2}$ times the length of the body. It is greyish with purple markings and dots and partially transparent. This is the largest British prawn and measures up to 4 inches in length.

COMMON PRAWN (*Leander squilla*)
Very common. It frequents the middle and lower shore and occurs in rock pools near the sea among the seaweeds. Similar in anatomy to the Common Prawn (*Leander serratus*) and edible, but smaller. The body is a pale brownish and almost transparent. It measures up to $2\frac{1}{2}$ inches in length.

HIPPOLYTE PRIDEAUXIANA
Fairly common on south and south-west coasts. It frequents the lower shore in pools among the seaweeds. Similar in anatomy to its relative the Chameleon Prawn, but larger and usually has a pale stripe, normally white, on its back. It measures up to $1\frac{1}{2}$ inches in length.

PALAEMONETES VARIANS Fig. 42 (d)
Locally common. It frequents estuaries and salt-marshes, where it occurs in brackish water, pools, drainage dykes, streams and similar. The body is almost transparent, with brownish-orange spots and markings. It measures up to $1\frac{1}{2}$ inches in length.

SPIRONTOCARIS CRANCHI
Locally common, but only on south and south-west coasts. It frequents the lower shore in pools among the seaweeds. Similar in anatomy to its relative Chameleon Prawn (*H. varians*). The body colour is variable and may be pale lemon with darker markings. It measures up to $\frac{3}{4}$ inch in length.

SHRIMPS
COMMON EDIBLE SHRIMP (*Crangon vulgaris*) Fig. 43 (b)
Very common. It frequents the lower shore, nearly always in sandy areas, bays, estuaries, and pools, sometimes seasonally occurring in immense numbers and supporting a local shrimp fishing industry, where it has been caught for centuries in nets from small boats, by trawl from horse and cart, or by the shrimper pushing his wide net in front of him in shallow water. The Common Shrimp is so familiar in fishmongers' shops that it scarcely needs description. The outer antennae is the same length as the body, while the carapace is flattened. In colour it is grey or dark brownish with darker or red markings, but even so, moving quickly over a sandy background, into which it also burrows, it is difficult to see. It can also change colour to blend with the background even more. It measures up to 3 inches in length.

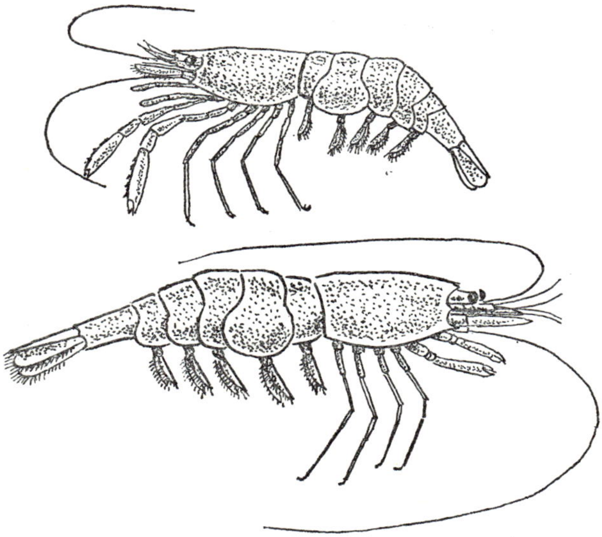

Fig. 43 Shrimps (a) (top) Hooded Shrimp *Athanas nitescens*
(b) (bottom) Common Shrimp *Crangon vulgaris*

HOODED SHRIMP (*Athanas nitescens*) Fig. 43 (a)
Locally common on south and west coasts of England. It frequents the lower shore where it occurs in pools among the seaweeds. The outer antennae is approximately half the length of the body, the first pair of legs being armed with large nippers, the tips of which may be white. Its shape makes it look like a miniature lobster. The body is red with a conspicuous white stripe running the full length of the back. It measures up to 1 inch in length.

LOBSTERS
COMMON LOBSTER (*Homarus vulgaris*) Fig. 44
Common. It frequents rocky areas offshore and is usually found hiding among rocks or in crevices in a pool. If discovered alive and trapped in a pool do not try to pick it up, as the formidable nippers seizing your fingers can cause painful injury. The anatomical shape of the Lobster seen on fishmongers' slabs is so familiar it needs no description, except that it is reddish due to being cooked. Alive it is dark blue. Lobsters can be long-lived marine creatures. It measures up to 20 inches in length, but elderly examples may be larger.

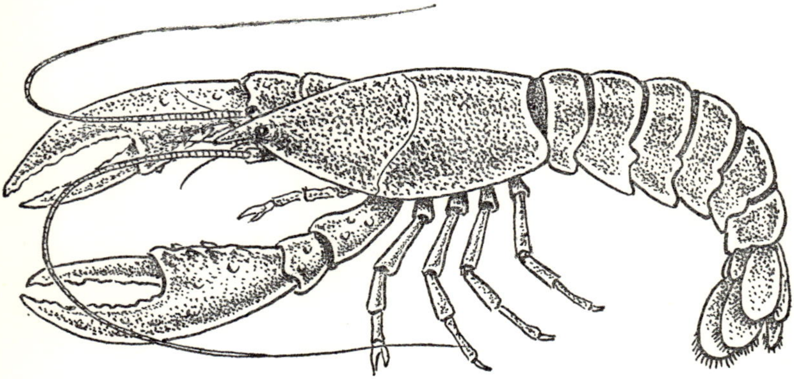

Fig. 44 Common Lobster *Homarus vulgaris*

SPINY LOBSTER, ROCK LOBSTER OR CRAWFISH (*Palinurus vulgaris*) Fig. 45

Fairly common on south and west coasts. It frequents offshore deep water, but may also be found in pools among the rocks. Similar to the Lobster, except that it has numerous tough spines, with antennae the length of the body, but it does not have the large formidable nippers. It is brown in colour and is sometimes larger than the Lobster, measuring up to 24 inches.

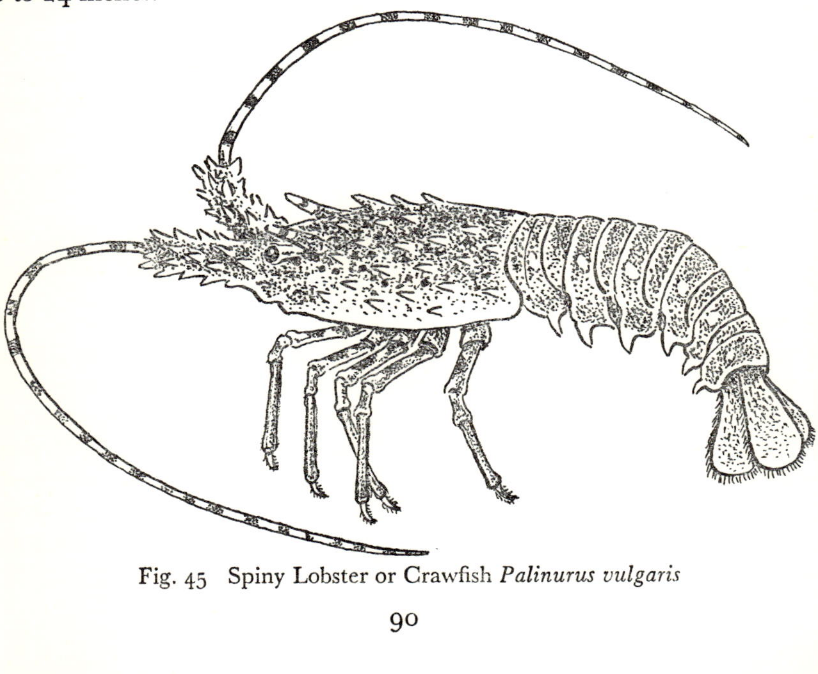

Fig. 45 Spiny Lobster or Crawfish *Palinurus vulgaris*

The threatening attitude of a Shore Crab *Carcinus maenas*.

A Common Starfish *Asterias rubens*, normally five-armed and this specimen may have lost one limb accidentally or through attack on it by a predator, but occasionally four- or six-armed specimens occur.

A Mute Swan with plumage in rather a dirty condition.

Oyster-catcher's eggs laid on shingle. They often blend remarkably against their background and so avoid detection.

SPINOUS SQUAT LOBSTER (*Galathea strigosa*)
Fairly common. It frequents the lower shore where it occurs under stones and rocks, also in pools under seaweed. Origin of name the same as the following species *Galathea squamifera.* The paler nippers and some of the legs are spiny. In colour it is bright red with crossing blue lines and dot marks, and measures up to 6 inches.

SQUAT LOBSTER (*Galmthea squamifera*)
Common. It frequents the lower shore where it occurs under rocks and stones, also in pools under seaweed. The name was given due to its habit of tucking its abdomen or tail part out of sight under its body. It is a brownish-greenish colour, sometimes with small red markings. It measures up to 3 inches.

The *Mysidacea* or Opossum Shrimp, so-named due to its shrimp-like appearance, differs from the true Shrimps by having eight pairs of swimming legs which are not suitable for crawling or burrowing. On the underside of its body it has a brood pouch, created from the overlapping plates, containing the eggs, from which hatch the young, as tiny Opossum Shrimps, which grow protected in the pouch until mature enough to leave. The large eyes are stalked and noticeably prominent, sometimes black. Usually the Opossum Shrimps occur in large shoals in their habitat.

CHAMELEON SHRIMP (*Praunus flexuosus*) Fig. 46
Very common. It frequents the lower shore, at the sea's edge, near rocks or in rock pools among various seaweeds, also found in flat, sandy estuaries and mudflats. Its colour varies considerably and may be of a near-colourless transparency to a dark blackish-grey, although brown or green are frequent colours. Sometimes it alters its colour to blend with its background if among seaweed. It measures up to 1 inch.

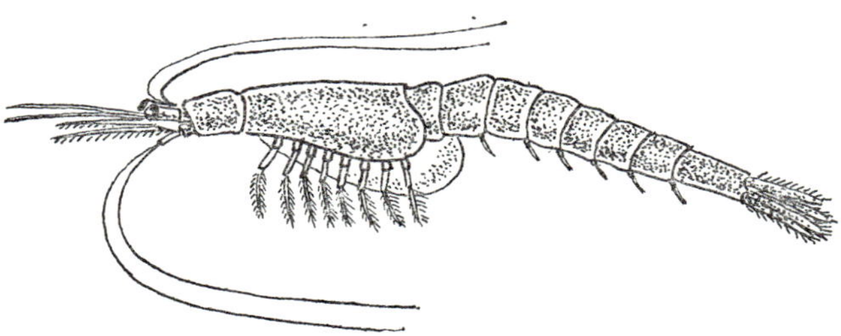

Fig. 46 Chameleon Shrimp *Praunus flexuosus*

GHOST SHRIMP (*Schistomysis spiritus*)
Common. It frequents the lower shore at the sea's edge and sometimes occurs in pools. Apart from the bold, black eyes it is entirely transparent and colourless, hence its name. It measures up to $\frac{3}{4}$ inch.

MIDGE SHRIMP (*Hemimsis lamornae*)
Locally common. It frequents rock pools staying near sheltering rocks, stones and seaweed. It is red or orange, with bold black eyes, and measures up to $\frac{1}{2}$ inch.

SANDHOPPERS

The *Amphipoda* or Sandhoppers may be mistaken for land insects because of their appearance. The body is curved from head to tail, with flattened sides, and has five pairs of walking legs and three pairs of appendages or swimmerets for swimming. The tail has three pairs of stiff appendages or uropods, the latter being used to thrust the animal forwards or off the ground when jumping. When revealed in their habitat they may walk into the nearest cover, jump or "hop", by suddenly straightening their bent body, or wriggle sideways, depending on the species. Despite their name only a few examples either hop or live in or on sand, as they also occur under decaying seaweed, shore debris, dead marine creatures, stones, rocks, and similar objects, in mud and muddy sand. They are extremely useful, being scavengers and clearing the seashore, estuaries, mudflats and coastline of vast quantities of decaying matter which would otherwise foul the area. There are a large number of members of the *Amphipoda*, some of which, due to their size or similarity, can be identified only by microscopic examination associated with consideration of their habitat and food and it has only been possible to refer to the more common or interesting examples which can be found and identified reasonably easily.

AMPHITHOË RUBRICATA Fig. 47 (a)
Common. It frequents all zones of stone and seaweed littered shore where it hides under these materials. Sometimes it also attaches together pieces of seaweed, using threads produced by its leg glands, to create a tubular hideout. In colour it is green, brownish-red or red and usually matches the seaweed it is among. The eyes are red and minute. It measures up to $\frac{1}{2}$ inch.

CHELURA TEREBRANS

Common. It is found in or upon submerged timber, jetty and pier piles, cast up driftwood and similar debris in association with the Gribble (*Limnoria lignorum*), *Chelura terebrans* apparently making use of the Gribble's borings in such wood. It is pale brown and measures up to $\frac{1}{3}$ inch.

COROPHIUM VOLUTATOR Fig. 47 (b)

Common. It frequents the middle shore in estuaries and mudflats where it creates colonies of short, U-shaped or straight tubes burrowed in soft mud or muddy sand. Evidence of these can be seen when the Sandhopper occurs in large numbers, by the way the surface of the mud is widely "perforated" with Corophium's tiny holes when the tide recedes.

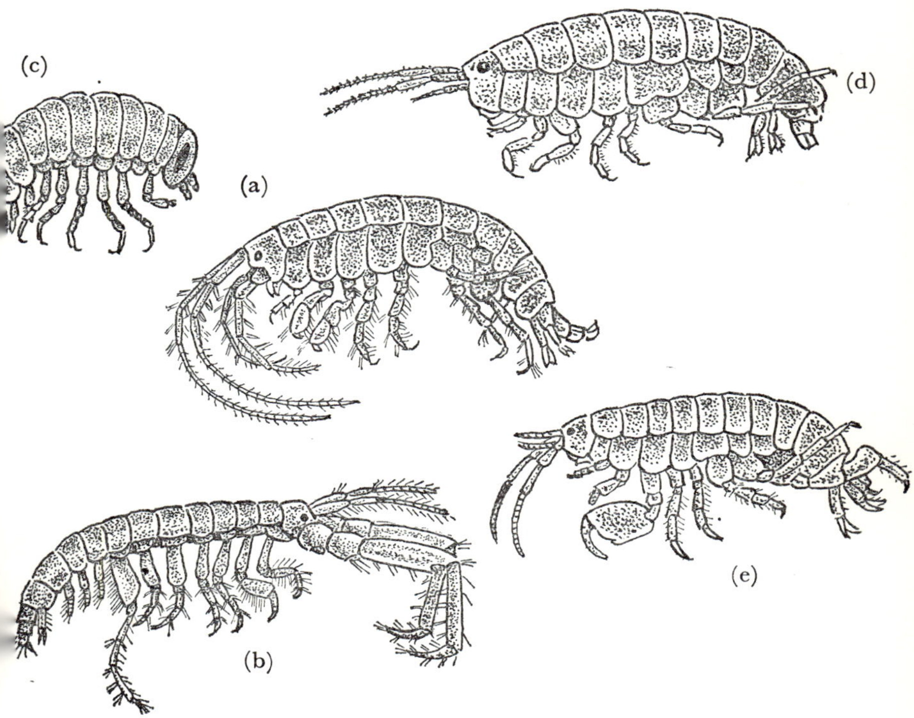

Fig. 47 Sandhoppers

:) *Hyperia galba* (b) *Corophium volutator*

(a) *Amphithoë rubricata*

1) *Talitrus saltator* (e) Shore Skipper *Orchestia gammarella*

When sea birds are seen wading upon the mud probing it with their beaks, they are probably feeding on a colony of these tiny creatures. It is a brownish-grey colour and measures $\frac{1}{8}$ inch in length.

GAMMARUS LOCUSTA
Very common. It frequents the middle and lower shore where it occurs under flattish stones and among seaweed, also brackish water. When revealed it normally wriggles on its side. It varies considerably in colour, but is usually a greenish or brownish with a pinkish area on each side of the body. The male measures up to 1 inch in length, the female is smaller.

HYPERIA GALBA Fig. 47 (c)
Common. It occurs under the bodies of jellyfish cast up on the shore It is sandy-brown colour, with red spots, and has green eyes, and measures up to $\frac{1}{2}$ inch.

SAND HOPPER (*Talitrus saltator*) Fig. 47 (d)
Very common. It frequents the upper shore where it burrows in th sand or under or among decaying seaweed and other debris. The Sand Hopper sometimes occurs in large numbers, hiding under their food - seaweed or dead animal jetsam – and when it is moved and they ar revealed they leap about in all directions. This species is in a transitor stage between being a marine animal which has almost left the sea an a land animal and are not now able to withstand long submersion i the sea. It is light brown and green or greyish-brown with a black lin running down the back. The male measures up to $\frac{3}{4}$ inch, the femal is slightly smaller

SHORE SKIPPER (*Orchestia gammarella*) Fig. 47 (e)
Common. It frequents all shore zones and is found in sand, shingl under stones and decaying seaweed. When revealed it normally jump It is reddish-brown or brownish-green and measures up to $\frac{3}{4}$ inch.

The *Isopoda,* commonly known as the Sea Slaters, occur on coast estuaries, seashores, in creeks and anywhere close to tidal sea water brackish water. By day they hide under rocks, stones, and similar, sand, or in crevices in cliffs, banks, seawalls, etc., emerging after du when the tide recedes to crawl to the shore and feed on vegetatio seaweeds, and animal matter. They are related to and in some cas closely resemble the familiar Woodlice or "Pill Bugs" occurring on lan behind tree bark, under stones, etc. Several of the Sea Slaters, like the Sandhoppers, have become almost independent of the sea and do not require it to reproduce, the female carrying the eggs and young about with her in a brood pouch until they have developed sufficiently to fend for themselves. The Isopods are sensitive to daylight, even bright moonlight, and will scuttle into the nearest hiding-place as soon as they are revealed. They are sought after, as prey, by certain crabs and wader sea birds. Some of the 18 species are extremely small, others rare, or closely identical to a related Slater, so that microscopic examination is necessary to determine the species and for this reason I have referred only to the larger examples or those most interesting. Their body is flattened dorso-ventrally with large outer antennae and seven pairs of legs.

GRIBBLE (*Limnoria lignorum*) Fig. 48
Very common. It frequents wooden pier and jetty piles, breakwaters and similar submerged harbour and shore installations; it may also be found in floating timber or wood that drifts on to the shore. A concentration of them will do serious damage by colonially boring their small tunnels a short distance into the wood. This causes the outer riddled wood layers to break away and allows the bared timber beneath also to be attacked continuously until it is useless and so eroded that the pile has to be replaced. The body has straight sides and rounded at each end. It measures up to $\frac{1}{5}$th inch in length.

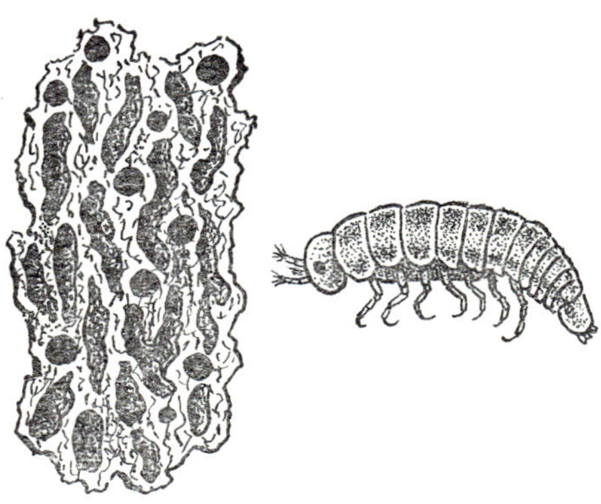

Fig. 48 Gribble *Limnoria lignorum*

IDOTEA NEGLECTA

Common. It frequents the lower shore where it occurs under seaweeds. The body is long and slender with the tail narrowing to a rounded point. It is brown in colour with whitish markings and measures up to 1 inch.

SEA SLATER (*Ligia oceanica*) Fig. 49 (a)

Very common. It frequents the upper shore of coasts and estuaries, where it occurs under rocks, stones, and similar, seaweed and debris; it also hides by day in chinks and cracks between rocks, seawall blocks, harbour walls and similar. When exposed in daylight it runs rapidly to the nearest suitable hiding place, but is active while feeding on the shore during darkness. The body is flat, oval in shape, and a greenish-brown or greenish-grey, with light spots and markings. It is one of the largest Isopods, measuring up to 1 inch, sometimes longer.

SPHAEROMA RUGICAUDA Fig. 49 (b)

Very common. It frequents the middle shore and also occurs in estuaries, harbours, brackish water and salt marsh pools. When disturbed, like the land Pill Bug or Woodlouse, it rolls itself into a ball. When it swims in the pools it does so on its back. The body is broad and oval in shape but blunt at each end. In colour it is whitish or greyish with some darker markings. It measures up to $\frac{1}{2}$ inch.

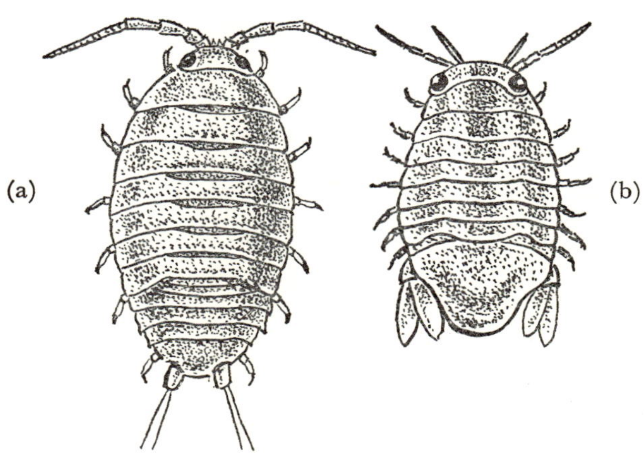

Fig. 49 (a) Sea Slater *Ligia oceanica*
(b) *Sphaeroma rugicauda*

Chapter 7

Molluscs – Univalves

The Molluscs, which include some of the familiar "shellfish", are a large group of soft-bodied animals, with either a single shell, two shells, or the shell may be minute and hidden or absent. The *"Mollusca"* as they are classed, from the Latin for "soft", include the Chitons (*Placophora*), the Univalves – Limpets, Top Shells, Periwinkles, Whelks and others (*Gastropoda*), the Bivalves – Mussels, Scallops, Oysters, Cockles, and others (*Lamellibranchia*) and Cuttlefish, Octopus and Squids (*Cephalopoda*).

Basically, the molluscs have a "foot", a muscular organ, on the underside of the animal's body, which is used either as a means of progression or for burrowing. A familiar example is the land mollusc the common garden snail. This "foot" continues, as part of the body, within the protective shell, combined with a fold of skin called the "mantle", the latter covering the "viscera" and forming the shell, which varies considerably in shape. The shell they inhabit is extended as the animal grows, and does not have to be shed periodically when the "shell" becomes too small to house the animal, as is the case with crabs and some crustacea. Between the "mantle" and body is the "mantle cavity" which usually contains the gills. On the upper side at the front of the body, *except in the bivalves,* is the "head" which contains tentacles and eyes. A "siphon" or "siphons" are also present in some species, such as the Whelk, which is used for respiration. The method of feeding, *except in the bivalves,* is by using the "radula", a horny ribbon which has rows of teeth. In the carnivores – flesh-eaters – the teeth are larger and fewer, the mouth being at the end of a proboscis, but in the vegetarians – plant-eaters – the teeth are small and numerous. Also upon the "foot" there may be a horny or limy "operculum" used as a lid to seal the aperture when the animal has need to withdraw into the shell. An easily-seen example is the Periwinkle.

Molluscs lay eggs that hatch into larvae which, in some species, live for a time as part of the plankton. In other species, the Sea Snails and Whelks, the young emerge from the egg capsules as miniatures of the adults.

This chapter covers only two sections of the *Mollusca,* the Chitons (*Placophora*) and the *Gastropoda* or Univalves, the latter being molluscs with a one-section or single shell. In the *Placophora* the simple body and foot is protected by eight overlapping shell plates. In the *Gastropoda* the Univalve shell is either of two shapes – cone-shaped, as in the Limpets, or coiled spirally, comprising several whorls, narrow and long or shorter and rounder, as in the Whelks, Periwinkles, and Top Shells, with considerable variations. The Univalve shell's entrance, out of which the animal's head partially projects, is called the "aperture".

THE CHITONS

The Chitons or Coat-of-Mail Shells (*Placophora*), like the Limpets, are perfectly adapted for their life upon the rocks and stones, but more primitive in construction. Their flattened, oval-shaped body is covered by eight shell plates on the back, surrounded by a "girdle". Beneath this there is a broad "foot", with gills and a small head, with the mouth in the middle, some species having small spines or tufts of minute bristles on the "girdle". The eight shell plates are jointed and overlap, so Chitons can grip rough surfaces. If washed off the rock the Chiton can curl itself into a tight ball to protect its body. In this way it sinks to the bottom and is bowled along by the wave action, but is unharmed, and as soon as it reaches a sheltered position it unrolls and climbs wherever it finds itself. It feeds on algae and occurs in slimier muddier situations than Limpets. Some of the Chitons are small and have dull coloration similar to the rocks, so they are inconspicuous and difficult to see.

ACANTHOCHITONA DISCREPANS

Uncommon. It frequents the lower shore. The girdle has spines and up to 20 tufts of bristles. The shell plates have a distinctive keel in the centre. In colour brownish. It measures up to ¾ inch in length.

BRISTLED COAT-OF-MAIL SHELL OR CHITON

(*Acanthochitona crinitus*) Fig. 50 (a)

Common on south and west coasts of England. It frequents the lowe

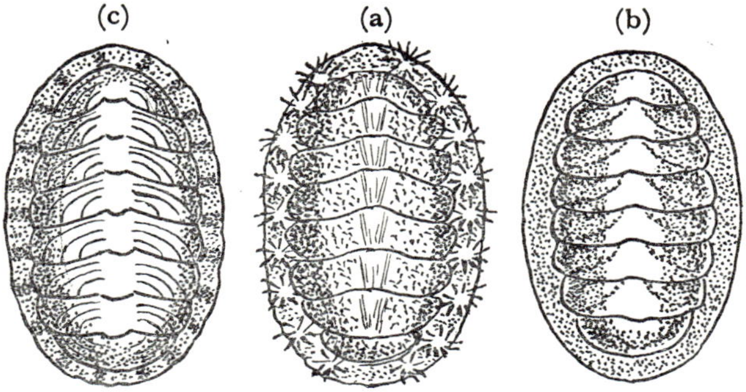

Fig. 50 (c) Red Chiton *Tonicella rubra*

(b) Grey Coat-of-Mail Shell *Lepidochitona cinereus*

(a) Bristled Coat-of-Mail Shell or Chiton *Acanthochitona crinitus*

shore. The girdle has spines and 18 tufts of bristles. The shell plates have a distinctive keel in the centre. In colour yellow, brown or reddish. It measures up to ½ inch.

GREY COAT-OF-MAIL SHELL (*Lepidochitona cinereus*) Fig. 50 (b)
The most common species. It frequents all shore levels, on rocky and tony shores. The broad girdle has large marginal spines but no bristle ufts. Its broad shell plates have a distinctive keel in the centre. Despite ts name it is coloured brown, red or green. It measures up to ½ inch.

RED CHITON (*Tonicella rubra*) Fig. 50 (c)
Common. It frequents the lower shore. The narrow girdle has no bristle tufts, but the shell plates have the distinctive keel in the centre. t is reddish or brownish and measures up to ½ inch.

HE LIMPETS

The Limpets (*Archaeogastropoda*) are easily distinguished from other Jnivalves by the conical shape of the shell, a miniature Chinese coolie's at. Beneath this the Limpet has a body similar to other members of ne group, with a "gill" or "gills", broad "foot", "viscera", "mantle", muscle", "head", "marginal tentacles".

"To cling like a limpet" is a description for anything gripping tightly, nd is one earned by this animal. By using its "foot" it is able to fix self almost immovably to a rock. This flat, round "foot" and the dhesion it creates, plus the shape of the broad-based shell, are able to resist the pounding of heavy seas, and the more forceful the water on the Limpet the more tightly it clings to its position by pulling the shell against the rock. A would-be predator finds it almost impossible to break this grip if the Limpet has been forewarned by a clumsy approach and is prepared for the attack. The secret of removing a Limpet undamaged lies in stealth and a sudden sideways blow when its shell is partly raised while moving or feeding.

The Limpet must retain some water around parts of its body to survive at low tide and to do this the shell must be watertight. Few rocks have exactly flat, smooth surfaces and to overcome this problem of fitting the shell base upon an irregular rock the Limpet uses one of two methods. It either wears away the edge of its shell to fit the uneven area of hard rock it has chosen as a "home site", or it files the softer rock itself to make an indentation to match the shell. If you examine Limpets on a rock you will notice how perfectly rock and shell edge fit together. The rock may also bear the ring-shaped indentation caused by the colonisation of earlier generations of Limpets.

Limpets leave their "home site" on the rock, etc., to crawl about, moving in a swaying motion with head and tentacles exposed, usually at high tide, in search of the algae it feeds on. As the tide recedes the Limpet moves back to its original position and so it must possess some kind of instinct to know its own "home indentation site" from the dozens of others, or perhaps it lays a trail or scent which can be retraced.

This animal begins life as an egg, freely and somewhat carelessly emitted into the sea, spawning usually taking place in winter. The tiny larvae which emerges swims with the plankton for a short time until it moves closer inshore to settle in the rocky habitat.

BLUE-RAYED, BLUE-SPOTTED or KINGFISHER LIMPET (*Patina pellucida*) Fig. 51 (a)

Common. It frequents the lower shore, where it occurs on oar-weed and other seaweeds, usually in areas where the fronds have been hollowed out during feeding. The shell is not perforated and there is no ledge inside it. It has a smooth surface and is semi-transparent. The apex is flat and closer to the front edge of the shell. An aid to identification in young examples are blue lines or rows of spots radiating from margin to apex, hence its names. These are not present on adults. It measures up to 1 inch.

COMMON LIMPET (*Patella vulgata*) Fig. 51 (b)

Very common. It frequents upper, middle and lower shores, where

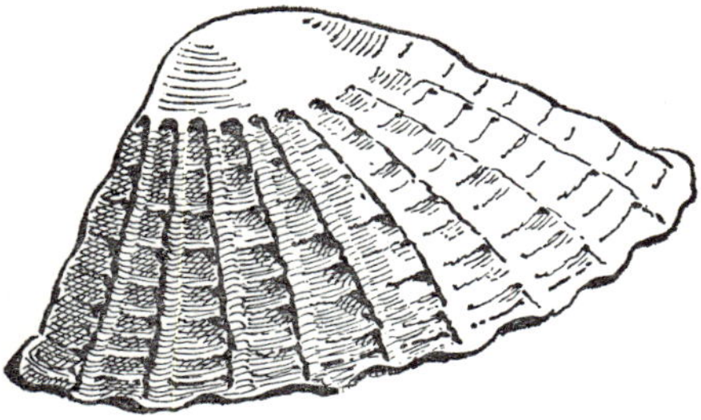

(b) Common Limpet *Patella vulgata*

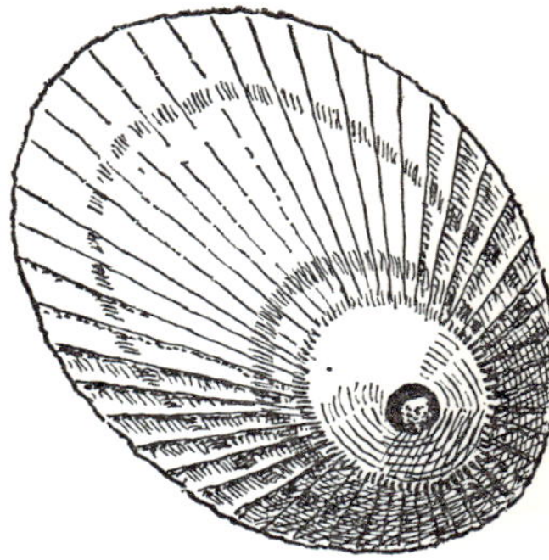

(c) Keyhole Limpet
Diodora apertura

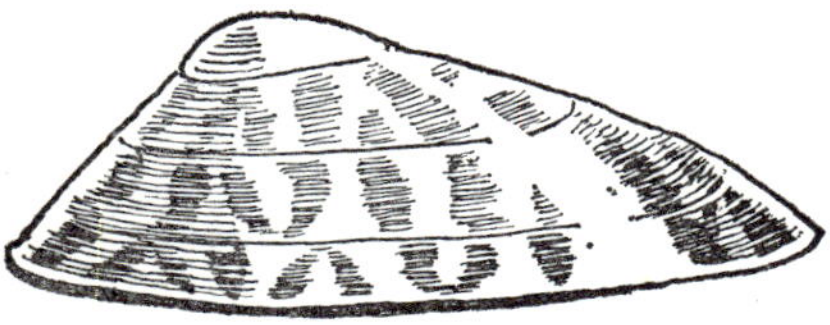

(f) Tortoiseshell Limpet
Acmaea testudinalis

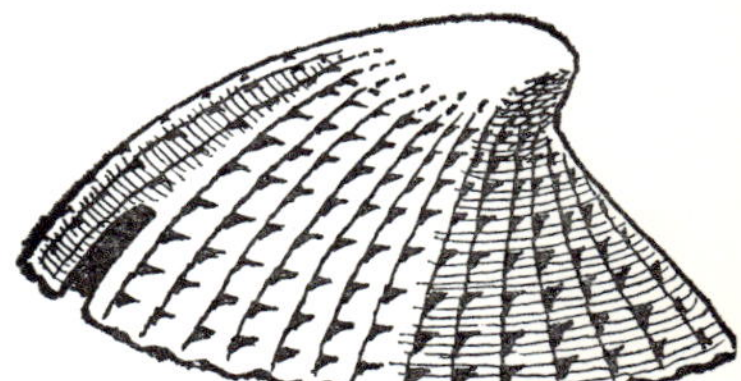

(e) Slit Limpet
Emarginula reticulata

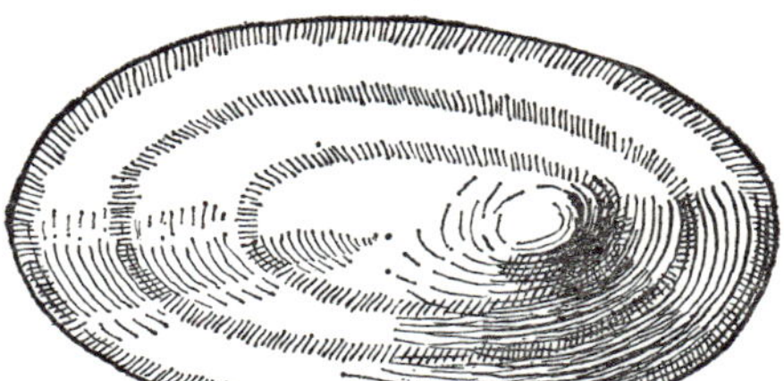

(g) White Tortoiseshell Limpet
Acmaea virginea

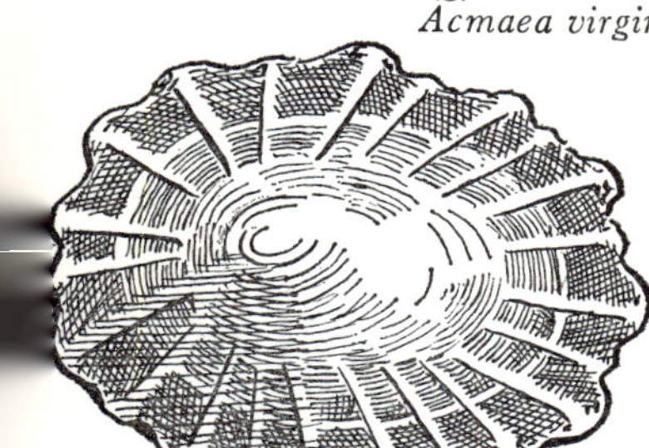

) *Patella depressa*

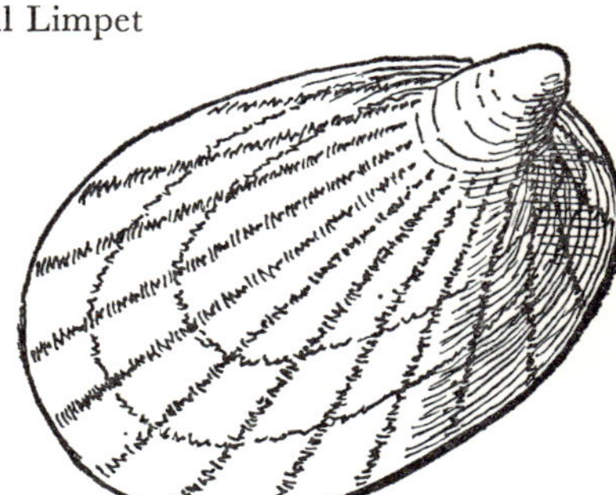

(a)

(a) Blue-rayed Limpet
Patina pellucida

Fig. 51 The Limpets

occurs on rocks, large stones, chalk boulders, etc., also in rock pools. The shell is not perforated, has no ledge inside shell or a curved apex. Its shell is tall, ribbed from margin to apex and rough surfaced, with the edge cut into even and rounded notches. Shell colour is variable, the inside surface white or yellow, the foot being greyish. It measures up to 2½ inches in length.

KEYHOLE LIMPET (*Diodora apertura*) Fig. 51 (c)
Fairly common, particularly on south and west coasts of England. It frequents the lower shore, on rocks. The shell has a single opening at the apex, used as part of their respiratory system. Shell radially ribbed from apex to edge. In colour brownish or yellowish with one or more darker radial patterns. It measures up to 1½ inches.

PATELLA ASPERSA
Uncommon, chiefly south-west and west coasts. It frequents the exposed lower shore, but is sometimes found in weedy rock pools on the upper shore. It is difficult to distinguish from the Common Limpet (*Patella vulgata*). The shell is flatter, not perforated, the interior having a transparent smooth surface. An aid to identification is the orange "foot". It measures up to 2½ inches.

PATELLA INTERMEDIA [DEPRESSA] Fig. 51 (d)
Uncommon. Found on the south-west coast, and possibly south coast of England. It frequents the exposed lower shore and may occasionally occur in weedy rock pools on the upper shore. Similar to *Patella vulgata* and *Patella aspersa*. The shell is flatter than *P. vulgata*, not perforated, ribbed, the interior being a dark colour, with darker areas on edge of exterior shell. The "foot" is also darker than in the Common Limpet. It measures up to 1½ inches.

SLIT LIMPET (*Emarginula reticulata*) Fig. 51 (e)
Fairly common. It frequents the lower shore. The aid to identification is the slit or notch in the front edge of the shell used as part of the respiratory system, the shell also being ribbed from apex to the edge the apex being fractionally curved. It is variably coloured, usually a greyish or brownish and measures up to ½ inch.

TORTOISESHELL LIMPET (*Acmaea testudinalis* [*tessulata*]) Fig. 51 (f)
Locally common in northern Britain only. It frequents the middle and

lower shore, in rock pools. The shell is not perforated, being smooth surfaced, the flattened apex positioned closer to the front edge of the shell. An aid to identification is the wavy, irregular pattern of reddish-brown and darker markings giving the exterior of the shell a resemblance to tortoiseshell; the shell interior is in darker colouring. It measures up to 1 inch.

WHITE TORTOISESHELL LIMPET (*Acmaea virginea*) Fig. 51 (g)

Common, more widespread than Tortoiseshell Limpet, also occurring on south and other coasts of Britain. It frequents the lower shore, among stones and oarweed, in rock pools. The shell is not perforated, being smooth surfaced and having a flattened apex close to the front edge of shell. It is whitish with pink or brown markings, the shell interior being lighter colouring. Smaller than Tortoiseshell Limpet, up to ¾ inch.

THE TOP SHELLS

The Top Shells (*Archaeogastropoda*) contain some of our most beautiful shell examples, particularly if they have been worn against the rocks to reveal the mother-of-pearl or nacreous layer under the shell surface. The shells are identified by their shape, broad at the base, near the rounded aperture, either conical, rounded or flattened. They are supposedly called Top Shells because of a resemblance to children's whip-tops. The animal, a sea snail, closes its shell aperture when it retreats within by means of a horny or limy, sometimes spirally patterned, "lid" or "operculum". When it is moving over the rock or seaweed surface the "foot", bearing "tentacles" on its sides, can be seen. Top Shells are usually found only in clean water, in pools and on similar surfaces of rocks, stones and seaweeds, never in muddy or sandy water or sand and mud shores and flats, the reason being that Top Shells have a primitive respiratory system and they are quickly killed by unclean, muddy, polluted water. They feed on seaweed and other marine plants.

Apart from the Common or Painted Top Shell, which lays its eggs in a long jelly-like string, the other Top Shell animals, spawning in spring and summer, emit the sperm and eggs into the sea where they are fertilised. The larvae hatch as minute crawling animals upon the sea bed or food plant and do not live prior to this among the plankton.

Empty shells of dead Top Shells species can sometimes be found higher up the shore than would be usual if alive, because the animal has died in its normal habitat in deeper water and then been cast ashore by the wave action.

COMMON, PAINTED OR LARGE TOP SHELL (*Calliostoma zizyphinum*) Fig. 52
Common. It frequents the lower shore. The shell comprises ten to twelve coils or whorls with a sharply pointed apex. It is pyramid shaped with straight sides and a flat base. In colour it is yellow, light pinkish-brown or a pink ground with red streaks, but in some areas white examples occur. The largest British Top Shell species, up to 1 inch in height and across.

Fig. 52 Common, Painted or Large Top Shell
Calliostoma zizyphinum

FLAT OR PURPLE TOP SHELL (*Gibbula umbilicalis*)
Locally common on south and west coasts. It frequents the lower and middle shore. The Shell usually comprises six or seven coils or whorls with a blunt apex. In colour it has a greenish or greyish ground with oblique, wide, purple stripes. Up to ½ inch in height, slightly more across.

GIBBULA MAGUS
Locally common on south and west coasts of England. It frequents the lower shore. The shell comprises eight coils or whorls, slightly flattened, and has a ridge around the base of the shell, also a blunt apex. In colour yellow with red or purple streaks. It is up to 1 inch in height, slightly wider.

GREY TOP SHELL OR "SILVER TOMMY" (*Gibbula cineraria*) Fig. 53
Very common. It frequents the middle and lower shore. The shell usually comprises seven coils or whorls, the lower whorls being flatter, with a

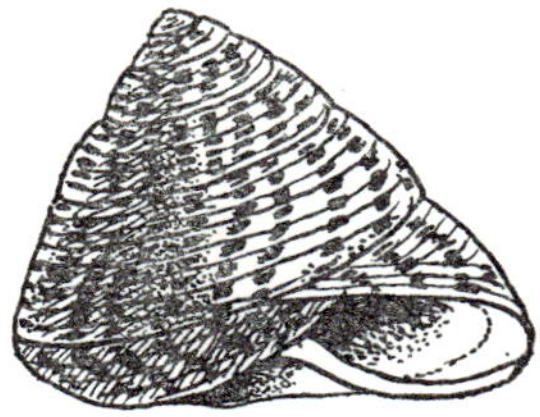

Fig. 53 Grey Top Shell *Gibbula cineraria*

blunt apex. The colour is pale grey ground with narrow, oblique, darker grey stripes. It measures up to ½ inch in height and across.

PHEASANT SHELL (*Tricolia pullus*)
Common on south, south-west and west coasts. It frequents the middle and lower shore where it occurs on the red seaweeds. The shell has several smooth, glossy coils or whorls, the last one, nearest the aperture, being over half the height of the shell. It is upright, much less wide, and unlike other Top Shells in shape, more closely resembling a Periwinkle. The colour is yellowish or buff with reddish or purplish line markings or spots. It measures up to ⅓ inch in height.

THICK TOP SHELL OR TOOTHED WINKLE (*Gibbula lineata*)
Common, but only on the south-west and west coasts as far northwards as North Wales. It frequents the rocky middle shore. The shell comprises six coils or whorls, thick, rounded in shape, with a blunt apex. In colour it is mottled greenish or greyish ground, with dark purple almost black, winding irregular streaks. The aid to identification is the mother-of-pearl aperture, also the yellow or silvery apex. It measures up to 1 inch in height and across.

THE PERIWINKLES
The Periwinkles (*Mesogastropoda*) are among the most common Sea Snails. They are found around our coasts, in estuaries and on the seashore, some species being able to exist successfully in creeks, inlets, on mudflats, close to salt marshes, on seawalls, almost on dry land in several cases, in widely variable salt water, where other shellfish are absent. As the tide recedes they retreat into pools, in crevices on breakwaters, or hide under stones, damp seaweed and other objects. Their shell is much tougher, to resist frequent rolling about by the waves, with fewer whorls, but is less attractive, usually being one or two colours, plain or streaked. Although they have a similarity, only the Common Periwinkle

is edible and sold as "winkles" on fishmongers' slabs and stalls. The Periwinkles are vegetarians and feed largely on seaweeds and lichens.

The female of all the species, with the exception of the Rough Periwinkle, lays eggs, after fertilisation by the male. These eggs are usually laid in jelly-like masses upon seaweeds. In some species, after hatching, the tiny larvae become members of the plankton in early life; in other species they stay on the shore weeds and rocks. Periwinkles are attacked by marine worms which destroy or damage some of their internal organs.

BANDED CHINK SHELL (*Lacuna vincta*)
Very common. It frequents the lower shore, among rocks and stones, seaweeds, particularly oar-weed. The shell, semi-transparent, comprises six coils or whorls and has a sharply pointed apex. Colour yellow or buffish-yellow patterned with reddish bands. It measures up to $\frac{1}{2}$ inch in height.

COMMON EDIBLE PERIWINKLE (*Littorina littorea*) Fig. 54 (a)
Very common, widespread around coast in bays and estuaries, gravelly and sandy shore, on mudflats, frequenting the upper, middle and lower shore, where it occurs on bare rocks and stones, chalk boulders, on seawalls, among seaweeds, on breakwaters, boat wrecks, pier piles, etc., also in pools. It is often found exposed and dormant at low water awaiting the return of the tide. The familiar edible winkle of fishmongers' shops and stalls. The shell is rounded, tough in construction, with only a few coils or whorls and a sharp pointed apex. It has a horny "lid" or operculum sealing the shell aperture. In colour dark grey, black, brown or reddish with darker concentric lines. The largest British species, measures up to 1 inch in height.

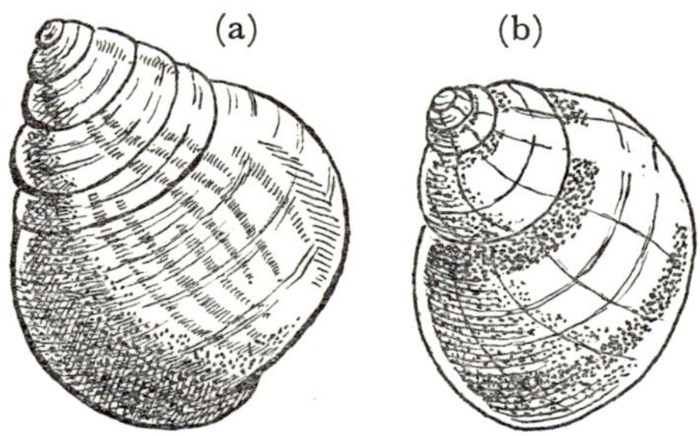

Fig. 54 (a) Common Periwinkle *Littorina littorea* (b) Rough Periwinkle *Littorina saxatilis*

FLAT PERIWINKLE (*Littorina littoralis*)
Very common. It frequents the middle and lower shore, on wrack seaweeds. The rounded shell has several coils or whorls, the aid to identification being that the apex has a flat top. It is very variable in colour, either red, brown, purple, orange, yellow, green or streaked with a majority of one of these colours. It measures up to $\frac{1}{2}$ inch in height, slightly more across.

ROUGH PERIWINKLE (*Littorina saxatilis*) Fig. 54 (b)
Very common. It frequents the upper and middle shore. The rounded shell is ribbed and comprises up to 9 coils or whorls and has a distinctively pointed apex. An aid to identification is the rough-feeling texture of the shell. The female gives birth to tiny shelled young. Very variable in colour, black, brown, red, yellow or white. It measures up to $\frac{1}{2}$ inch in height.

SMALL PERIWINKLE (*Littorina neritoides*)
Common. It frequents the upper shore and above high water level, occurring in protecting crevices in rocks, on the cliff faces, especially those facing south, and in pools, this species favouring more exposed situations than other species, sometimes almost on dry land or land only infrequently washed by spray. For this existence it is able to obtain oxygen from the air by means of a "lung", and is probably in a transitional stage of becoming a land mollusc and will eventually leave, or partially leave, the sea. The shell is long and smooth, with a pointed apex. In colour it is a dark brown or grey, an aid to identification being the shell has a "bloom" on the surface similar to that on plums, grapes and other fruits. It measures up to $\frac{1}{4}$ inch in height, less across.

THE WHELKS

The empty shells of the Whelks (*Neogastropoda*) are almost as familiar as the Periwinkles on the seashore. The thick shells are much larger, proportionately narrower, with pointed spire and apex, the aperture being generally large and oval-shaped. The Whelks have a siphon pointing upwards which is used to secure clean water for respiration from a level above its head instead of from a lower level where the mud is probably disturbed by the Whelk's progress on its "foot". Whelks are carnivorous, preying upon other living creatures and also scavengers, two species being attracted to carrion. They can be a pest by damaging fish caught in nets. The Common Whelk is itself eaten, by us, as it has been from ancient times. Whelks are also used as bait. Empty shells of the Common and Dog Whelks are frequently used during the life cycle of the Hermit Crab as protective quarters.

The Whelks begin life as an egg, usually laid by the females in a numerous mass – except the Thick-lipped Dog Whelk, which lays them separately – on rocks, stones and seaweeds. From these the young Whelks emerge and develop in stages. Following their emergence the vacant egg capsules sometimes break away and are cast ashore among debris.

COMMON WHELK OR BUCKIE (*Buccinum undatum*) Fig. 55
Common. It frequents the lower shore, among rocks, stones, gravel, sand and mud. The shell comprises up to eight deeply grooved whorls, a pointed apex and large oval aperture. In colour it is yellow or sandy-yellow. Empty shell examples are common but if they have been lying on the shore for any length of time they are usually white. It is the largest British inshore Univalve, measuring up to 6 inches in height. The egg capsules are normally laid in rounded masses upon rocks, but are commonly found on the middle and lower shore among debris as dried, straw-coloured, sponge-like balls or lumps.

Fig. 55 Common Whelk *Buccinum undatum*

DOG WHELK OR DOG WINKLE (*Nucella* [*Purpura*] *lapillus*
Fig. 56 (a)
Very common. It frequents the middle and lower shore, on rocks, stone and other areas where barnacles and mussels are attached and ofte found in large numbers. It also feeds on acorn barnacles, limpets, peri winkles, top shells and other sea snails. This is achieved in one way b

moving on top of the victim, such as a limpet or acorn barnacle, and laboriously slowly drilling a narrow hole in the victim's shell, then extending its protruding proboscis and rasp teeth into the prey's body and eating it. Alternatively, to feed on mussels or barnacles, it forces their plates or shell valves apart, perhaps by killing the victim with the poisonous purple dye, purpurin, which the Dog Whelk secretes. After the victim's muscles relax the Dog Whelk can easily extend its proboscis through the gap which it has made and get on with its meal. Evidence of the Dog Whelk's presence will be empty shells of barnacles and mussels with the tell-tale round holes bored in them. The shell varies in shape and colour, usually being cone-shaped with spiral ridges giving the shell an irregular outline and with a large oval aperture. Colour varies according to its food victims. If it mainly feeds on barnacles it is dull white, but when it preys on mussels it is darker, mauvish or dark brown, with brown bands. Some examples also have white and black whorls, like humbug sweets, or yellow-brown banded with dark brown, a mixed diet or change of diet from barnacles to mussels being a suggested cause of this. Where they are yellow or orange this is believed to be due to wave action, not diet. It measures up to $1\frac{1}{2}$ inches in height.

NETTED DOG WHELK (*Nassarius reticulatus*) Fig. 56 (b)
Common. It frequents the lower shore, where it occurs on sand and muddy sand, sometimes in sandy hollows while the tide is out. An aid to identification is that it feeds on carrion and several may be found under or upon the corpse of a dead fish or other sea creature. The shell comprises up to ten whorls, deeply grooved and ridged, giving the

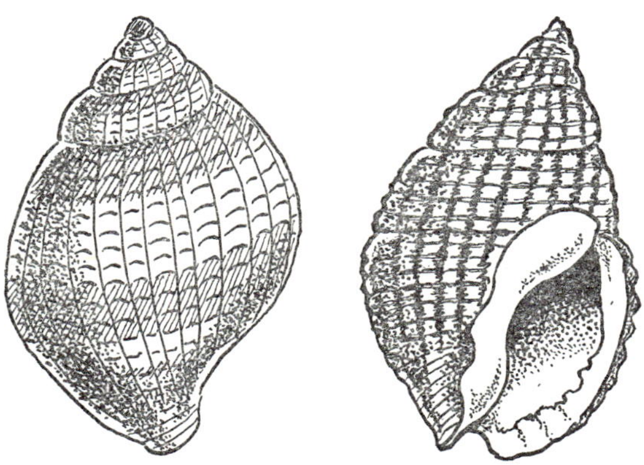

Fig. 56 (a) Dog Whelk *Nucella lapillus* (b) Netted Dog Whelk *Nassarius reticulatus*

shell a network appearance, hence its name, also a pointed apex. In colour it is orange-brown or pale brown. It measures up to 1½ inches in height.

OYSTER DRILL (*Urosalpinx cinerea*)
Frequents only the coasts of Kent and Essex, but is a serious pest where it does occur, primarily on or close to oyster beds. The shell comprises up to seven whorls with a pointed apex. In colour, brownish-grey. It was originally a North American species, but was accidentally introduced here with imported oysters. It feeds on young oysters and other bivalves after boring into the victim's shell. It measures up to 1½ inches in height.

STING WINKLE OR DRILL (*Ocenebra erinacea*)
Common on south-west coasts of England and Wales; occasional on the east coast. It frequents the lower shore, on rocks, muddy sand and gravel, but mainly on barnacle communities and oyster beds, where it is a pest. The shell comprises up to ten whorls, with spiral and longitudinal ridges giving it a rugged, irregular, corrugated appearance, and has a pointed apex. In colour dull white or pale grey with darker streaks. It feeds on bivalves, oysters and barnacles, after boring into the victim's shell. It measures up to 2 inches in height.

THICK-LIPPED DOG WHELK (*Nassarius incrassatus*)
Common. It frequents the lower shore, on rocks, stones and gravel, where it feeds on carrion. The shell comprises up to ten whorls, rounded, with longitudinal ridges and a pointed apex. In colour brown with darker bandings and has a brown area on the base. It measures up to ½ inch in height.

NECKLACE SHELLS, COWRIES, SPIRE SHELLS and VARIOUS OTHER UNIVALVES

This section of the Gastropods, Univalves, contains widely varying examples which can be frequently discovered, either on the seashore on coasts, in estuaries, on or near mudflats. They may occur alive in large numbers, or be found as empty shells.

CHINAMAN'S HAT (*Calyptraea chinensis*)
Locally common on south-west coasts of England. It frequents the lower shore, usually adhering limpet-like to stones, rocks or other shells. The shell is round with a low apex. An aid to identification is a conspicuousl

prominent ledge inside the shell. The colour is pinkish. It measures up to $1\frac{1}{4}$ inch in length.

COMMON NECKLACE SHELL (*Natica alderi*)
Common. It frequents the sandy lower shore, where it burrows. The shell comprises six whorls, rounded and shiny, also it has a large aperture. This Mesogastropod has a very large foot and when using it to crawl it covers much of the shell. Colour of the shell is brownish-yellow with darker or red markings or spots in spiral rows. It preys on bivalves, particularly *Tellina,* by boring a round hole in their shell, using an acid secreted by a gland, through which it inserts its proboscis to inject a digestive juice which partly digests the victim's soft flesh before the Necklace Shell sucks it out. It gets its name from the spiral coil or collar of jelly, in which it lays its eggs, that becomes encrusted with grains of sand. It measures up to $\frac{3}{4}$ inch in height and across.

EUROPEAN COWRIE (*Trivia monacha*)
Common. It frequents the lower shore, usually on rocks, where it feeds on compound sea squirts, using a long proboscis. Its empty shells are sometimes washed ashore. The shell, with a long slit as the aperture on the underside, has up to twenty-five ribs and is white in colour, the shell also having three purple-brown spots. The brightly coloured orange mantle tissues may conceal the shell when the animal is on the move, projecting its head and foot. It measures up to $\frac{1}{2}$ inch in length.

LAVER SPIRE SHELL (*Hydrobia ulvae*)
Very common. It frequents estuaries, mudflats, salt marshes and brackish water areas, feeding on marine vegetation, seaweeds, particularly sea lettuce. Sometimes occurs together in huge numbers. The shell comprises six whorls, elongated, the aperture having a pointed top. It measures up to $\frac{1}{4}$ inch in height.

NEEDLE SHELL (*Bittium reticulatum*) Fig. 57 (b)
Fairly common. It frequents the lower shore, where it occurs on dead seaweed and other debris. Sometimes in large numbers. The shell is a tall, pointed spire, comprising up to sixteen whorls. In colour red-brown. It measures up to $\frac{1}{2}$ inch in height and about $\frac{1}{8}$ inch across.

PELICAN'S FOOT SHELL (*Aporrhais pes-pelecani*) Fig. 58
Very common. It frequents the lower shore, where it burrows in muddy sand and gravel. The empty shells are frequently cast ashore. The shell

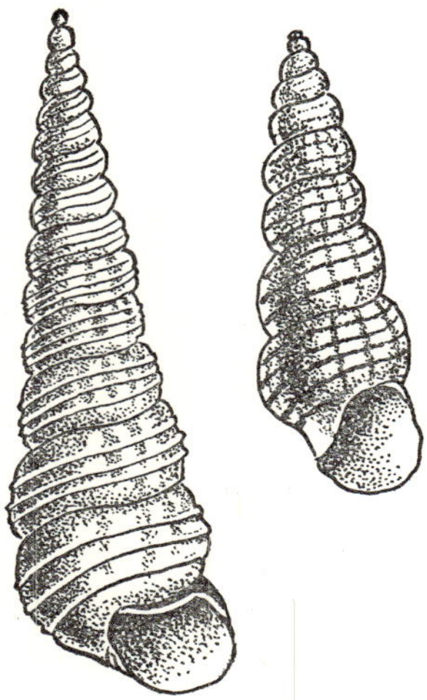

Fig. 57 (a) Tower Shell *Turritella communis* (b) Needle Shell *Bittium reticulatum*

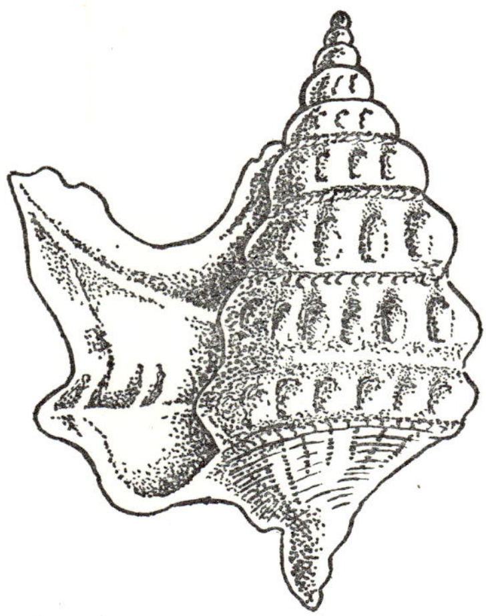

Fig. 58 Pelican's Foot Shell *Aporrhais pes-pelecani*

has several ornamented whorls. An aid to identification, but only in the adults, is the expanded outer part of the aperture with divisions giving the resemblance to a pelican's foot. In colour brownish. It measures up to $1\frac{3}{4}$ inches in length.

SLIPPER LIMPET (*Crepidula fornicata*) Fig. 59

Frequents east, south-east, south, south-west and west coasts of England where it is a serious pest in oyster beds; also in estuaries and creeks. The shell consists of a chain of up to twelve individual animals, one animal attaching itself to the back of another, male and female together, the males changing sex as they develop. The apex of each shell is shaped like a Cap of Liberty. Also, the shell has a ledge on the inside which gives it a supposed resemblance to a rounded slipper. It was accidentally introduced into England from North America with imported oysters in 1880. It is very prolific but sluggish, sometimes occurring in masses attaching itself to oysters and mussels which it does not attack, but it smothers or starves them of food by rivalling them for the microscopic plankton organisms. It is not related to the true shore limpets. In colour brownish, it measures up to 1½ inches in length.

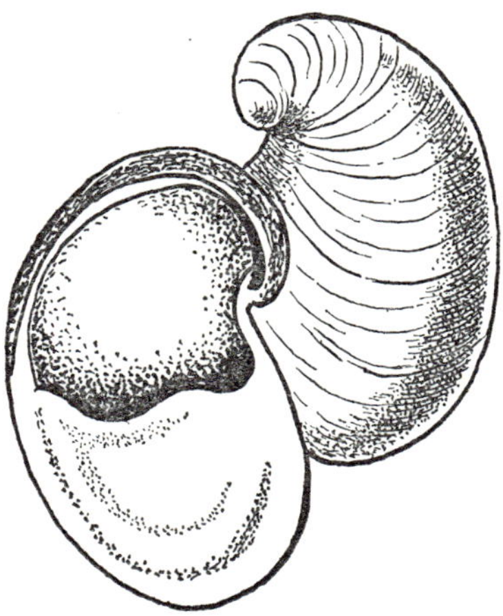

Fig. 59 Slipper Limpet *Crepidula fornicata*

TOWER SHELL (*Turritella communis*) Fig. 57 (a)

Very common. It frequents the lower shore, burrowing in muddy sand and gravel. Empty shells are frequently washed ashore. The long, narrow, tapering shell has up to nineteen whorls, some of them ridged, also a pointed apex and round aperture. It measures up to 2¼ inches in length.

Chapter 8

Molluscs – Bivalves

The Bivalve Molluscs (*Lamellibranchia*) differ from the Univalve Molluscs in having two valves, not one, as their shell, between which their compressed body is protectively enclosed. These two valves are usually hinged and have an elastic-like ligament to unite them. The valves or shells are closed when required by the Bivalve Mollusc contracting two powerful muscles. When the muscles are relaxed the elastic ligament separates or opens the valves. At the top or close to the edge of each valve is a raised protuberance or umbo.

The Bivalves do not have a head as such, although they have a mouth and some species, for example the Scallops, have eyes upon their body mantle. They feed and breathe by obtaining a current of water, sometimes through very long, tubular mantle extensions called the siphons, one for inhaling, the other for exhaling, these being either separate or joined. This water enters the large gills, which separates the microscopic food particles from it and passes these to the mouth, while the unwanted water is expelled through the exhalant siphon. The Bivalve Molluscs are either suspension feeders, such as the Cockles, which obtain microscopic food particles, suspended in the sea water, or are deposit feeders obtaining their organic food lying as a deposit on the surface of the sand, mud, etc.

Most of the Bivalve Molluscs, like the Univalve Molluscs, have a flattened "foot" extended and withdrawn between the two valves, used for moving about, burrowing in sand or mud and, as in the Piddocks, gripping the side of the excavated tunnel. The "foot" also produces a liquid which forms threads by which some of the Bivalves attach themselves to rocks and other objects.

OYSTERS

The Oysters (*Ostreidae*) have been relished here as a delicacy since the days of the Ancient Britons. The Romans exported vast numbers of native Oysters from the beds in southern England to be eaten by gourmets in Rome.

Since then the Oyster as a health-giving morsel has remained popular, but the population of the native English Oyster (*Ostrea edulis*) has declined drastically and a number of near-disasters on beds of cultivated Oysters has meant the necessary importation of young or "seed" Oysters from Holland and France to supplement our own stock. The American Oyster (*Ostrea virginica*) and Portuguese Oyster (*Ostrea angulata*) have been introduced and "planted", for the same reason, in beds, though it is claimed by food connoisseurs that neither of these has the superior flavour of our Native Oyster. Due to the water in our beds being lower in temperature the larger, long, narrow, fluted Portuguese Oyster spawns only very occasionally and the American Oyster has not established itself.

The Oyster's flattened body is contained within hinged inequivalve shells, one of the two valves being larger than the other. A powerful muscle is used by contraction to tightly close the two valves or shell. It has a "foot" for only a short period in its early life. It feeds, as well as breathes, by a current of water being continuously drawn through the gills which filter the microscopic food and pass it to the mouth before expelling the water.

Strange as it may seem, the Oyster begins its life as a male, then alters its sex to female, following this by producing eggs or sperm in alternating spawning periods. This is quite a common occurrence with Bivalve Molluscs.

The Oyster reproduces by discharging tiny larvae or "spat". Sometimes an individual Oyster may produce several hundred thousand during the spawning period, usually from May to September. Each embryo contains a minute larvae, inside a tiny but formed shell and has cilia or vibrating threads used to swim freely. After a short period in this planktonic stage the larvae seeks a place as a habitat where, in company with many others, it attaches itself to a stone, rock or shell or other fixed object, by its lower valve or shell. This it does by producing a cement from a gland in the "foot". After this the foot becomes shrunken and weak. However, the Oysters may become dislodged from their "home" at some time in their life and are then free, lying upon the habitat or embedded in mud. In the Oyster beds laid and specially cultivated by private companies clean shell known as "cultch", is spread to produce a suitable habitat for the "spat" to utilize after their planktonic stage. It is certain that but for the skilled culture of oysters by private, sometimes ancient, companies this bivalve delicacy would have long since disappeared from our coasts. The Thames Estuary is the principal area of oyster cultivation in England. The food supply the oysters can obtain here gives the native oysters the green colour and delicate flavour which makes them world-renowned amongst oyster gourmets.

The Saddle oysters (*Anomiidae*) could be mistaken for smaller examples of native oysters, unless several factors are remembered. Saddle Oysters have the oval hole in the lower or under-valve through which limy threads pass to attach them to a stone or rock. They adhere firmly to the object all their lives, unless removed by some means, and as they grow so the approximately round shell shape changes on its margins to follow the alignment of the object it is fixed to. When found upon the shells of other Molluscs it has been said to have a supposed resemblance to the saddle on a horse's back and so was given the name of "Saddle Oyster".

EDIBLE, FLAT OR NATIVE OYSTER (*Ostrea edulis*)

Common, particularly in the south-east of England, but their number fluctuates due to the depredations of predators, pollution, or effects of severe winters. It frequents the shallow water of estuaries, bays and tidal creeks, where it sometimes occurs in huge numbers and creates the well-known "beds". In similar habitats it is also cultivated artificially as a food delicacy. It occurs singly in deep water. The empty shells are frequently found on the lower shore, some of them, on their exterior, having acted as host for Acorn Barnacles, have attached tubes of marine worms, or been bored with the small, round holes of the Sponge, Cliona celata. The thick, flat shell is rounded in shape, with irregular edges. It is whitish or greyish in colour, rough on the exterior, with irregular layers, which are the growth lines formed when the shell was added to as the Oyster grew. Inside, the shell is smooth and whitish, with a silvery pearl-like lustre. The Oyster is sometimes attached to stones, rocks and other objects, with its left valve or lower shell on the underside by means of a cement, not threads as used by the Saddle Oysters. The shell measures up to 4 inches in diameter, but specimens 6 inches or more are occasionally found.

RIBBED SADDLE OYSTER (*Monia patelliformis*)

Common, but more frequently found alive on rocks, stones and shells in deeper water of lower shore, where it adheres itself by the same method as the Saddle Oyster (*Anomia ephippium*). Empty shells of valves may be found higher up the shore. The shell is rounded and flat, with numbers of "ribs" or ridges almost encircling the upper shell from the raised protuberance or umbo near one edge. Its exterior colour is greyish or greenish-white with rusty-brown edging or markings. It measures up to 1 inch in diameter.

SADDLE OYSTER OR SILVER SHELL (*Anomia ephippium*)
Fig. 60
Common. It frequents the middle and lower shore, where it is found on or under rocks, stones, shells and similar objects, also seaweed. It attaches itself tightly to these, not by a cement, as does the EDIBLE OYSTER, but by means of limy threads from a hole in the underside valve or shell. The frail shell is roundish and flat, with a raised protuberance or umbo near one edge. Its exterior colour is a dirty white, yellow, pinkish or brown, but inside it is usually a glossy silvery-white. It measures up to 2 inches in diameter; examples usually found are about 1 inch to 1½ inches.

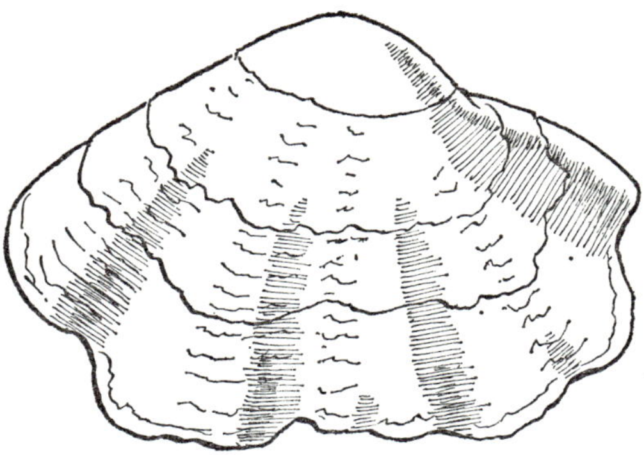

Fig. 60 Saddle Oyster *Anomia ephippium*

MUSSELS

Some species of the Mussels (*Mytilidae*) are familiar Bivalves in creeks, stony mudflats, estuaries and on our coast and seashore, where they occur in large "beds". The shell is usually a triangular-shape, with an umbo or raised protuberance on the upper surface or at one end. This streamline shape has the purpose of giving least resistance to the force of the waves. It is attached securely to the rock, stone, pier piles, jetty or other object by a number of "threads" called the byssus. The "threads" are produced from a gland in the "foot", which manufactures a liquid that on emission hardens quickly to form a thread. A number of these are sited in all directions by the "foot". This allows the Mussel shell to swing as to face the moving tide with its pointed end and reduce the possibility of its being dislodged. Should these "threads" be broken by some other means the Mussel will produce more. Sometimes the "threads" are attached to a stone or rock in mud which becomes more deeply embedded. To avoid being dragged down the Mussel can sever the threads and move about on its "foot" until it finds a more suitable object and re-attach itself.

The two valves of the shell are hinged, virtually the same size and fit perfectly together. They are kept closed by a large, powerful muscle. Mussels feed by partially opening the valves, drawing in a current of water which is strained by the gills to extract microscopic food particles from it, these being passed to the small mouth, while the water and waste products are expelled.

Mussels reproduce by spawning, emitting eggs or sperm into the water. The young larvae live as plankton for a time, then they descend to a suitable site where they move about actively on their "foot" until they produce the threads to attach themselves to the object where they may spend much or all their life.

The outside of Mussel shells often acts as host sites for Acorn Barnacles. This bivalve is attacked by Dog Whelks and Starfish, which force open the valves to feed on the succulent victim inside. Tiny Pea Crabs also inveigle their way inside the Mussel shell, usually Common and Horse Mussels, probably when the Mussel valves are ajar while it is drawing in water. Therein the crabs not only steal the Mussel's food particles for themselves but even feed on their host's gill tissues. Unfortunately for the crab it sometimes stays as an unwanted guest too long and grows too large, so that it is unable to escape by the way it entered. Thus it remains imprisoned, its only hope of escape perhaps being that its host will die and the powerful muscles be relaxed so that the two valves will come apart, or that a starfish will come along and break open the "gaol" so that the Pea Crab can leap out quickly. Mussels create havoc in an oyster bed, particularly when they are assembled in masses. The attaching "threads" of the numerous Mussels causes the deposition of mud and this, coupled with the weight of the Mussels, smothers and kills the oysters.

The COMMON MUSSEL is edible and a human food, but does not have quite the attraction of oysters or cockles. The reason for this may be that Mussels more rapidly absorb pollution, particularly from untreated sewage, than do other edible bivalves and in the past there have been cases of people who ate them being seriously ill through the Mussel's contamination, although the Mussels are not apparently affected. There is no risk of infection or cause for concern if the Mussels have been gathered from unpolluted beds or have been kept in purifying tanks to cleanse them. Mussels are also harvested in vast numbers to use as fish bait. The crushed shell also has a commercial use.

BEAN HORSE MUSSEL (*Modiolus phaseolinus*)
Common. It frequents the lower shore, where it occurs on rocks and stones. The blunt-pointed shell has a raised protuberance on the upper surface and also has "thorns" over much of its surface. Aid to identification is that when these are taken off the shell will be seen to be bean-shaped, hence its name. Sometimes the "threads" that attach it to the habitat also have clusters of shell fragments on them. The shell exterior colour is yellowish-purple and yellowish-brown. It measures up to $\frac{3}{4}$ inch in length.

BEARDED HORSE MUSSEL (*Modiolus barbatus*) Fig. 61 (a)
Common, particularly on south and west coasts of England. It frequents the lower shore where it occurs on rocks and stony places. The blunt-pointed shell has a raised protuberance or umbo on its upper surface.

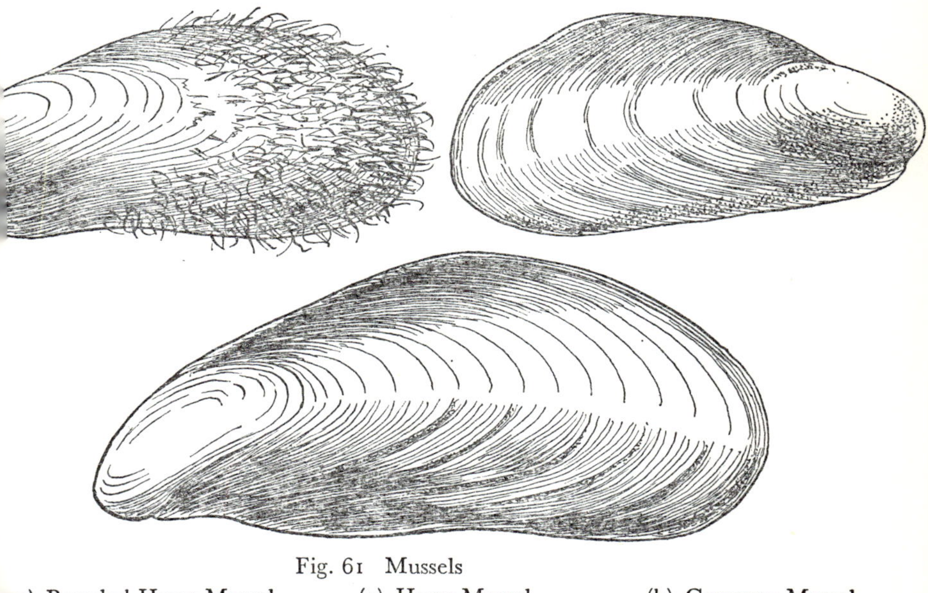

Fig. 61 Mussels

a) Bearded Horse Mussel *Modiolus barbatus*
(c) Horse Mussel *Modiolus modiolus*
(b) Common Mussel *Mytilus edulis*

It also bears a fringe of thick, barbed prolongations like thorns, giving it its name. The colour of the shell exterior is yellow or purplish-red and yellowish-brown. It measures up to 2 inches.

COMMON OR EDIBLE MUSSEL (*Mytilus edulis*) Fig. 61 (b)
Very common. It frequents the middle and lower shore, on coasts, mud-flats and in estuaries, where it occurs in large numbers clustered on rocks, stones, breakwaters, pier piles, jetties, wrecks, buoys, and other objects. It is regularly exposed at low tide, and also found in rock pools. The empty shell valves are common in these habitats. The pointed shell has a raised protuberance or umbo at its tip. It varies in shape and colour, the exterior shell being a dark blue almost blackish-blue, sometimes greenish, brownish or purple, a variety *pallida* being a yellowish-straw colour. The interior is pearly white. It measures up to 4 inches.

FAN MUSSEL, FAN SHELL OR PINNA (*Pinna fragilis*)
[*Pinnidae*]
Locally common. It frequents the lower shore, but is more common in deep water, where it occurs singly in muddy, sandy, gravelly areas of sheltered bays. It buries itself vertically in these habitats with the pointed end downwards and the wider end of the shell protruding above the surface. It attaches itself to deeply-buried rocks and stones by means of its long, tough, strong threads. The flattened shell shape is like a half-closed fan, hence its name. The colour of the shell exterior is pale yellow-brown. It is the largest British bivalve; it measures up to 12 inches in length and 6 inches at its widest part, but occasionally larger and wider examples are found.

GREEN CRENELLA (*Musculus discors*)
Common. It frequents the middle and lower shore, where several may be found together in rock pools among seaweeds or on the open shore with seaweeds. The shell is not so triangular-shaped as other Mussels but more oval and it has wide grooves or ribs from the raised protuberance or umbo at one end. The colour of the shell exterior is bright green or brownish-green, but where this is worn away it may be yellow-brown. It measures up to ½ inch in length.

HORSE MUSSEL (*Modiolus modiolus*) Fig. 61 (c)
Common. It frequents offshore deep water, but sometimes it is cast ashore attached to broad-leaved seaweeds. Young, yellowish-brown examples, bearing shell and stone fragments, may be found in rock

pools on the lower shore. The thick, blunt pointed shell, with a horny fringe on the edge, has a raised protuberance or umbo on its upper surface. The colour of the shell exterior is a dark purplish-brown or blackish-brown. This is the largest British Mussel, measuring up to 6 inches.

MARBLED CRENELLA (*Musculus marmoratus*)
Common. It frequents the lower shore where it occurs in rock pools, among Sea Squirts or attached to seaweeds, shells, etc. The shell is not so triangular-shaped as other Mussels, but more oval or rounded. It has a raised protuberance or umbo at one end. The colour of the shell exterior is pale green with darker green, yellowish, purplish or brown-red markings. It measures up to ¾ inch.

SCALLOPS

The familiar shape of the shell of the Scallops (*Pectinidae*) has been used for centuries in heraldic arms, as a design for pilgrims' badges, copied for fabrics, as a pattern for trinket boxes, spoon bowls, etc. For much longer it has also been an article of food, the Great or Edible Scallop, also known as the Clam, and the Queen Scallop both being fished for in our deeper coastal waters.

The Scallop's simple, flattened body is contained within a hinged, rounded, equivalve shell, except in the case of the Great or Edible Scallop which is inequivalve. The underside shell or valve is convex with the surface curving outwards. There is a good reason for this. Scallops mostly lie with their valves in a horizontal position upon the bed or object, thus the rounded surface of the valve or shell lifts the animal clear of the sand, gravel or mud bottom when its valves are open and it is respiring or feeding. If the shell were flat its margins would be on or in the sand or mud and the Scallop would not secure undisturbed water. The two valves are held together by an elastic-like ligament, as a hinge extending along the "ears" or "wings" part of the shells. The convex surface of each shell or valve also has an umbo or raised protuberance where the narrowing point of the shell joins the "ears" or "wings". The valves are kept closed when required by one large, powerful, central muscle. If a Scallop can be seen feeding it will be noticed that inside the margin of the shell there is a row of tiny tentacles in-between which are numerous shining points which are its eyes. The Scallops are fixed by threads near the hinge when very young but some species free themselves to move unrestricted when adult animals.

EDIBLE SCALLOP, GREAT SCALLOP OR CLAM (*Pecten maximus*) Fig. 62 (a)

Very common. It frequents deep water, but sometimes examples occur on the lower shore. Empty shells or valves are often cast ashore. The rigid shells are nearly round in shape, but inequivalve, one larger than the other. The upper valve is almost flat, while the lower or underside valve is convex, with the surface curving outwards. It is attached only by the threads to objects when small, later becoming free and swimming by a rapid flapping motion with its valves, with the rounded open side foremost and the hinge pointing rearwards. If it is attacked or becomes alarmed it can return the way it has come or take an avoiding route by expelling a jet of water from the gills to move suddenly in a different direction but now with the hinge pointing forwards. This action, sometimes in a jerky progression, also applies to other species of Scallop which swim freely and are not attached. The exterior colour of the shell is yellowish, grey, brown, or a reddish-brown. It is the largest British Scallop, measuring up to 5 inches in diameter.

HUNCHBACK SCALLOP (*Chlamys distorta*) Fig. 62 (b)

Common. It frequents the lower shore and deep water, especially o rocky coasts and attached to oarweed. Empty valves or shells are some times cast ashore. The hinged shell is roughly rounded in shape, bot shells or valves being the same size or equivalve. Each shell is als convex, with the surface curving outwards, but becomes irregular o distorted through its life when spent in rock crevices, giving it a hunch backed appearance. It has up to 70 raised, sharp, narrow, large an small "ribs". This Scallop while small attaches itself by threads but late cements its underside shell or valve to an object. The exterior colou of the shell is white, yellowish, brownish, or red, mottled with streal and markings. It measures up to 1½ inches in diameter.

QUEEN SCALLOP (*Chlamys opercularis*) Fig. 62 (c)

Common. It frequents deep water, particularly on sandy coasts, whe it swims vigorously in shoals, but sometimes occurs on the lower shor Empty shells or valves are cast ashore. The hinged shells are round shape and both valves are the same size or equivalve. Both shells a also convex and have up to 20 wide, rounded "ribs". The exteri shell colour is very varied, from white to purples and browns, sometim spotted or blotched; pinkish with rose-red markings is a very beautif example. It measures up to 3 inches in diameter.

A colony of Star Sea Squirts *Botryllus schlosseri* encrusted upon a pebble giving the latter a floral appearance.

A motley gathering of Acorn Barnacles, a Common Limpet, several Common Periwinkles, a fragment of a Dog Whelk and two Grey Top Shells.

The beauty of some pebble beaches can be judged by this close-up of several examples.

A typical pebble beach with, although beautiful, virtually no marine life.

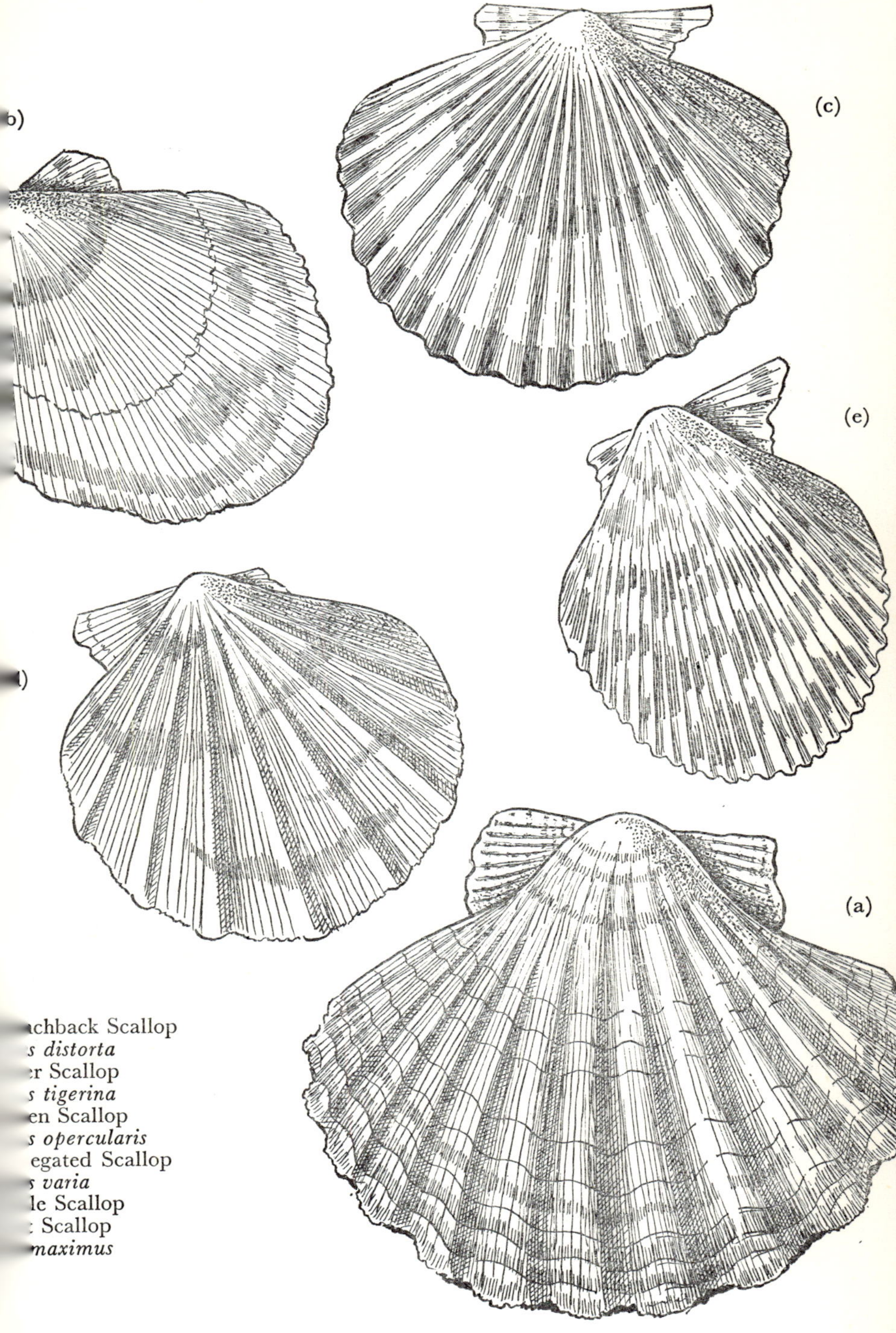

ıchback Scallop
s distorta
ɾr Scallop
s tigerina
en Scallop
s opercularis
egated Scallop
s varia
le Scallop
: Scallop
maximus

Fig. 62 The Scallops

TIGER SCALLOP (*Chlamys tigerina*) Fig. 62 (d)
Common. It frequents deep water, but sometimes occurs on the lower shore. Empty shells or valves are cast ashore. The hinged shells are round in shape and both are equivalve or the same size, except that the "ears" or "wings" are unequal in size.

Both shells are also convex, with the surface curving outwards, with numerous radiating and broad ribs. It is attached by threads to an object when young, but later is free and able to swim and spends most of its existence among stones, in crevices in rocks and similar habitats. The exterior colour of the shell is very variable, yellowish, brownish, purples, occasionally whitish, but has patternings of spots and stripes (hence its name) which are an aid to identification. It measures up to 1 inch in diameter.

VARIEGATED SCALLOP (*Chlamys varia*) Fig. 62 (e)
Common. It frequents the lower shore and deep water. The empty shells or valves are occasionally cast ashore. The hinged shells are more oval than round in shape and both valves or shells are equivalve, or the same size, but the "ears" or "wings" are unequal. Both shells are also convex with the surface curving outwards and have up to 30 "ribs". This Scallop swims freely, but examples also attach themselves by threads. The exterior colour of the shell is very varied and may be a single colour, such as a yellow, brown, red, pink, purple, rarely white, or streaked and mottled with different colours. It measures up to $1\frac{1}{2}$ inches in diameter.

COCKLES

The Cockles (*Cardiidae*) are also so familiar that they need littl description. The shells are approximately globular in shape, each havin a curving umbo or raised protuberance at one end, close to the hinge the shell also having numerous "ribs". The latter may serve the purpos of helping the Cockle "grip" the sand or muddy sand in which it i usually buried. It buries itself only a short distance below the surfac and projects its two short siphons above the surface to draw in wate with one, to breathe and also extract the minute plant organisms in th plankton on which it feeds, after which the water is expelled via th other siphon. When this takes place as the tide is draining off the san where the Cockles are living the tiny spurts of water, known a "pissing", give the Cockle gatherer a clue where each Cockle is situate The Cockle also has a very large, powerful "foot" which is used t burrow in the sand. It also can be used to roll the shell over the surface of the sand or leap above it in the hope of escape if exposed by an enemy. The Cockle achieves this movement by bending its fleshy "foot" in the niddle and pressing it against the sand, then suddenly straightening he "foot" so that it operates like a released spring and rolls the Cockle along the sand or may even fling it a considerable distance in the air f enough thrust is obtained.

Cockles are collected either by gatherers digging or scraping them rom the sand, sometimes using special sickle-shaped scraping tools or akes with large teeth. In some areas the gatherers walk the sandy beds arefooted to make the Cockles come to the surface, though the opposite ffect might be expected. In other areas a wooden instrument called "jumbo" is used upon the sand to achieve the same purpose. After gathering the Cockles are either boiled or steamed and then go to market. Due to the cleaner habitats where they live Cockles are less likely to be ffected by pollution when gathered in clean, sandy beds.

Apart from man, Cockles have several natural enemies. Starfish will rey on them, so do several snails. Flatfish lying on the sandy beds eat he young ones. At low tide they are vulnerable to being dug out by ulls, which have a way of opening them. They simply carry their apture to any rocks that are nearby and drop the Cockle. On hitting he rock it is shattered open and the gull obtains the delicacy.

OMMON OR EDIBLE COCKLE (*Cardium edule*) Fig. 63 (a)
Very common. It frequents the middle and lower shore, where it occurs uried in muddy sand and clean sand, in estuaries, bays, and suitable eashores, also brackish water. Sometimes it occurs in vast numbers as beds". The empty shells are also very commonly found on the shore. The hinged shell is globular, curving outwards, with numerous, probably ver two dozen, obvious, radiating "ribs". The exterior colour of the hell is creamy-white, yellowish or light brown. When living in brackish ater the exterior of the shell sometimes has a bluish tinge. The Edible Cockle has a greater popularity as a sea food than other Molluscs, with erhaps the exception of the Common Periwinkle probably because they e more edibly accessible than Winkles and cheaper than Oysters, so at a visit to the seaside is not complete without a dish of Cockles and negar from a sea food stall. The shell is also crushed and has a number commercial purposes. This Cockle measures up to 2 inches in length.

ITTLE COCKLE (*Cardium exiguum*)
ommon. It frequents muddy sand in deeper water offshore, but some-

times occurs on the lower shore buried in a similar habitat; it also occur in estuaries. The hinged shell is similar to the EDIBLE COCKLE, bu much smaller, with up to 20 "ribs". Colour of exterior of the shell i dull brownish-white. It measures up to ½ inch.

PRICKLY COCKLE (*Cardium echinatum*) Fig. 63 (b)
Common. It frequents deep water, but occasionally occurs on the lowe shore, usually in sand, sometimes muddy sand. The empty shells ar occasionally cast ashore. The hinged shell is thicker than other Cockle with about 18 obvious "ribs". An aid to identification is that the exterio shell surface has numerous backward-pointing, short, connected, curve spines. Colour of the shell exterior is whitish or light brownish. I measures up to 2½ inches.

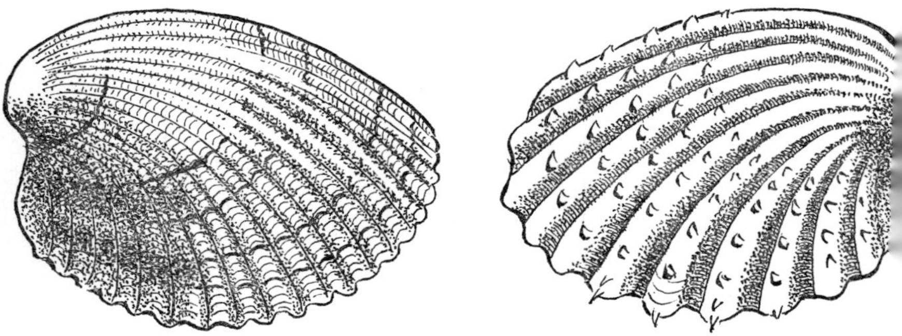

Fig. 63 (a) Common Edible Cockle *Cardium edule* (b) Prickly Cockle *Cardium echinatum*

SPINY COCKLE OR RED-NOSED COCKLE (*Cardium aculeatu*
Locally common, but only on south coast of England. It frequents de water where it occurs buried in sand. Occasionally empty shells may cast up on the lower shore. The thin, hinged shell has about 20 "ribs with numerous backward-pointing, tough, curved spines. This Cock gets one of its names from the bright red colour of its "foot". The colo of the shell exterior is light greyish-brown. It is the largest British Cock measuring up to 3 inches.

RAZOR SHELLS

The Razor Shells (*Solenidae*) are unmistakable due to their lo and narrow valves, each of which supposedly has a resemblance to t blade of an old-fashioned "cut-throat" razor. They have a hinge at one end, outside the valves, which gape, and are thin, smooth and very fragile. On the lower part of the shore the single valves, or broken parts of them, are a common sight. The living animal itself will be buried in the sand, coming to the surface only to breathe and feed through its projecting siphons when the tide returns. It is extremely sensitive to vibration and withdraws into the sand as soon as it detects anything, such as human feet, approaching. To dig one out a slow and careful approach is needed to where it is positioned below a depression in the sand.

Then a quick plunge with the spade may catch it by surprise, but equally it may be too speedy and by means of its large, powerful "foot" it will dig itself downwards into the sand to escape quicker than you can shovel away the sand in pursuit. If one is dug up it should be placed in a tin or box for examination, because if laid flat on the sand its "foot" will bend downwards and as soon as it gets a purchase it will haul its valves into an upright position, then immediately and rapidly dig itself out of sight.

ENSIS ARCUATUS Fig. 64 (c)
Common. It frequents the lower shore where occurs in sand. The shell is slightly curved with each end truncated. The "foot" is creamy-white. The exterior colour of the shell is yellow-brown with a violet-brown area. It measures up to 6 inches in length.

GROOVED RAZOR SHELL (*Solen marginatus*) Fig. 64 (b)
Locally common, chiefly on south and west coasts of England. It frequents the lower shore, where it occurs in muddy sand. The shell is straight and equal throughout its length, with a deep indentation at the anterior end. The "foot" is dull red in colour. The exterior colour of the shell or valves is light yellowish-brown, red-brown or orange. It measures up to 6 inches.

POD RAZOR (*Ensis siliqua*) Fig. 64 (d)
Common. It frequents the lower shore where it occurs in clean sand. The shell or valves are straight with each end truncated. The "foot" is creamy-white. The exterior colour of the shell is, dark yellowish-brown with a violet-brown area. The largest British Razor Shell, it measures up to 8 inches.

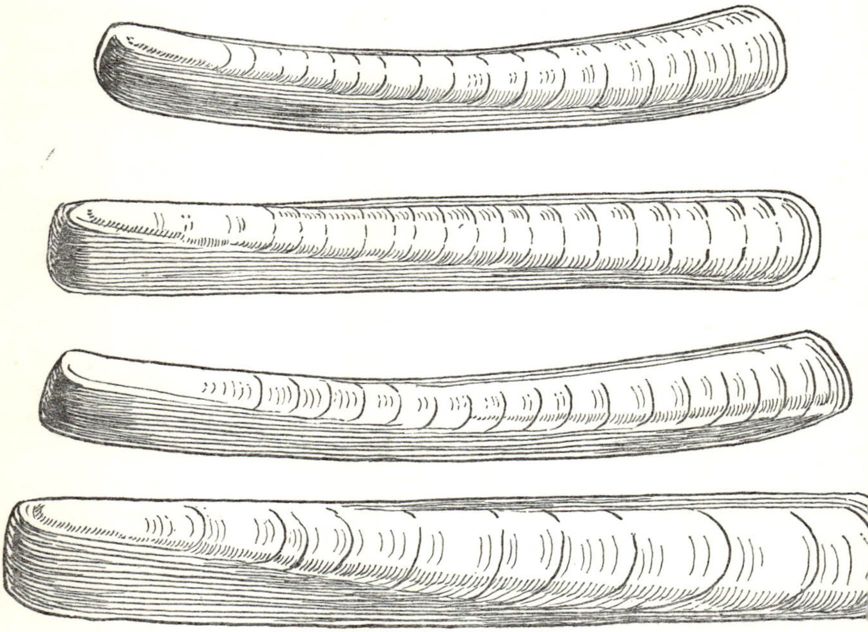

Fig. 64 The Razor Shells
(a) Sword Razor *Ensis ensis*
(b) Grooved Razor *Solen marginatus*
(c) *Ensis arcuatus*
(d) Pod Razor *Ensis siliqua*

SWORD RAZOR (*Ensis ensis*) Fig. 64 (a)
Common. It frequents the lower shore where it occurs in sand. Tl shell has a pronounced curve with the anterior end round, while tl posterior end tapers. The "foot" is pale red-brown. The exterior colo of the shell is yellowish with a greyish-white area the length of one sid It measures up to 4 inches.

ROCK AND WOOD BORING BIVALVES

Several of the Bivalve Molluscs can bore their way into various roc slate, chalk, clay, peat, etc., while others can bore into wood, such submerged timbers. This is done to protect themselves from enemi For the same reason other Bivalves burrow into sand or mud.

The Rock Borers, chief of which are the Piddocks (*Pholadid* achieve this by alternately contracting the two muscles, normally us to close the valves, and rasping the toothed edge of the shell to and fro against the rock, chalk or clay to make a penetration. The circular hole is made by the Bivalve changing its position as it bores. Sometimes you will find lumps of rock, limestone, or chalk on the shore which are riddled right through with the borings of Piddocks.

The shell of the Piddocks is elongated or oval and cut away so that the rounded "foot" protrudes through the gape and can be used to grip the bored tunnel.

The Wood Boring Bivalves are the infamous Teredos or Shipworms (*Teredinidae*) which have caused immense damage to and destruction of wooden ships, wood pier piles, jetties, dykes, underwater constructions and immersed timbers for centuries. The Shipworms look more like worms than shelled Bivalves because the shell has been evolved into a tiny, highly efficient boring "tool" at the end of the long, otherwise naked, body. Due to the size of their lengthy body Shipworms are unable to withdraw into their shells. The round "foot" is used as a gripping sucker to maintain the cutting shells or valves against the surface of the wood where the boring is taking place.

In a similar boring method to the Piddocks one of the two powerful muscles, normally used in other Bivalves to close the two valves, contracts and causes the sharp teeth on the valves to scrape the surface of the wood. By continually repeating this, coupled with muscular control and release of other areas of the valves, with the "foot" also regularly changing its position, a smooth, round tunnel is laboriously bored into the timber. The two separate siphons, at the tunnel entrance, which inhale and exhale water for food and respiration, have two limy "pallets" at the posterior end, which are used, when the siphons are withdrawn, to close the entrance of the bored tunnel. Shipworms feed upon a proportion of the wood as they bore into it and less so upon microscopic particles in the water.

If a badly infected piece of timber attacked by Shipworms and removed for replacement is examined it will be found to be honeycombed with tunnels and yet none of the tunnels internally opens into another tunnel. In some way the Shipworms are able to know they are close to a fellow Shipworm's tunnel and take an avoiding route. When they are unable to burrow Shipworms will cover the interior of their tunnel with a limy layer and continue to live in the habitat until they die naturally, usually after a year or so of existence. Various methods have been tried through the centuries, since the days of the Greeks, Egyptians and Romans, to protect the timbers of wooden ships, using lead and copper plates, charring the hull exterior, coating it with powdered glass, "soaps", pitch and various paints, to deter the Shipworms. Research is still continuing into the problem.

ROCK-BORING BIVALVES

COMMON PIDDOCK (*Pholas dactylus*) Fig. 65

Locally common on south and south-west coasts of England. It frequents the lower shore where it bores into chalk, red sandstone, shale, sunken timber and wood and peat that is submerged. Sometimes the burrows are 12 inches or more in length. The shell is elongated, bearing up to 50 longitudinal rows of spines where "ribs" and ridges meet. Colour of the shell is white, but may be stained by the habitat. This is the largest British Piddock, the shell measuring up to 6 inches in length, but 3 to 4 inches is more usual.

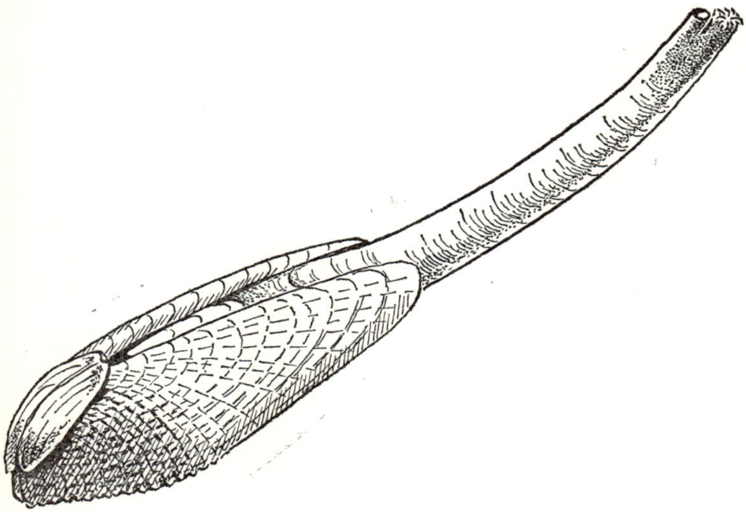

Fig. 65 Common Piddock *Pholas dactylus*

LITTLE PIDDOCK (*Barnea parva*)

Locally common. It bores into red sandstone, sometimes limestone o clay. The shell is more oval oblong, with a cut-away shaped appearanc in the front and has numerous scale-like ridges. The colour of the she is white when clean, but may be stained red or brownish-yellow throug burrowing into the habitat. It measures up to 2 inches.

PAPER PIDDOCK (*Pholadidea loscombiana*)

Uncommon, chiefly found on south and south-west coasts of Englan

It frequents the lower shore where it bores into clay and red sandstone. The shell is partly rounded, with an obvious furrow running diagonally from the umbones or raised protuberances to the edge of the shell. The upper surface is also ridged. At the end of its shell is a trumpet-like or cup-like membranous extension which protects the posterior base of the siphons and is an aid to identification. When it has reached the adult stage of $1\frac{1}{2}$ inches this Piddock stops boring burrows into rock and, having prepared for itself a suitable burrow it remains protected within, apparently satisfied to feed through its siphons and in season reproduce. The inactivity causes the powerful "foot" to become atrophied and the anterior opening through which the "foot" was extended when in use is closed to a small hole by the development of tissues covered with a horny material. The shell is white. It measures up to $1\frac{1}{2}$ inches.

RED-NOSE (*Hiatella striata*) [*Hiatellidae*]
Very common. It frequents the lower shore where it bores deep round burrows into limestone, chalk and soft rocks. It does not bore into hard rocks, but on these it attaches itself to the surface by means of its threads fixed in the crevices. Sometimes it also attaches itself to mussels in beds or to the large seaweeds such as oarweed. The shell is approximately oblong and irregular. Its colour is dull white, but it gets its name because the two siphons have red tips and these sometimes project, supposedly like a red nose, from the habitat burrow. Not a Piddock. It measures up to 1 inch.

WRINKLED ROCK BORER (*Hiatella arctica*) [*Hiatellidae*] Fig. 66
Very common. It frequents the lower shore where it bores deep round burrows into limestone, chalk and soft rocks. It does not bore into hard rocks, but attaches itself to the surface of these by means of its threads in the crevices; it also attaches to large seaweeds such as oarweeds. The thick shell is approximately oblong and irregular in shape and has a somewhat wrinkled appearance due to the numerous ridges. The colour is dull white. Not a Piddock. See RED-NOSE. It measures up to $\frac{3}{4}$ inch.

WOOD BORING BIVALVES
SHIPWORM (*Teredo navalis*)
Common. It bores into wood piles of piers, jetties, breakwaters, wooden vessels, floating and submerged timbers. Length up to 6 inches.

SHIPWORM (*Teredo norvegica*) Fig. 67
Locally common. It bores into wood piles of piers, jetties, breakwaters,

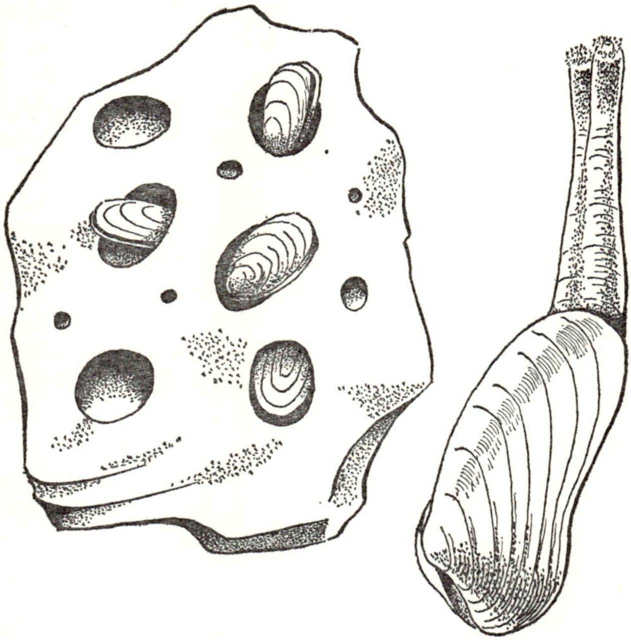

Fig. 66 Wrinkled Rock Borer *Hiatella arctica*

wooden vessel hulls, floating and submerged timbers. The largest British Shipworm, up to 12 inches in length.

SHIPWORM (*Teredo megotara*)
Not so common as the other two Shipworms. It bores into floating woo debris usually. It has an ear-like projection on the shell.

WOOD PIDDOCK (*Xylophaga dorsalis*) [*Xylophaginidae*]
It bores short spherical tunnels in floating wood, usually about the sam diameter as its shell, but does not feed on the timber while it does thi The shell is similar to that of the Shipworm's but entirely encloses th body of the Wood Piddock which can withdraw inside, excepting th siphon, when it needs to do so. It does not attach itself to the burrow it makes, nor does it have limy pallets to close the bored tunnel. I measures up to $\frac{3}{4}$ inch in diameter.

VARIOUS BIVALVE SHELLS
A visitor to the coast and seashore, sometimes also in estuaries, wi

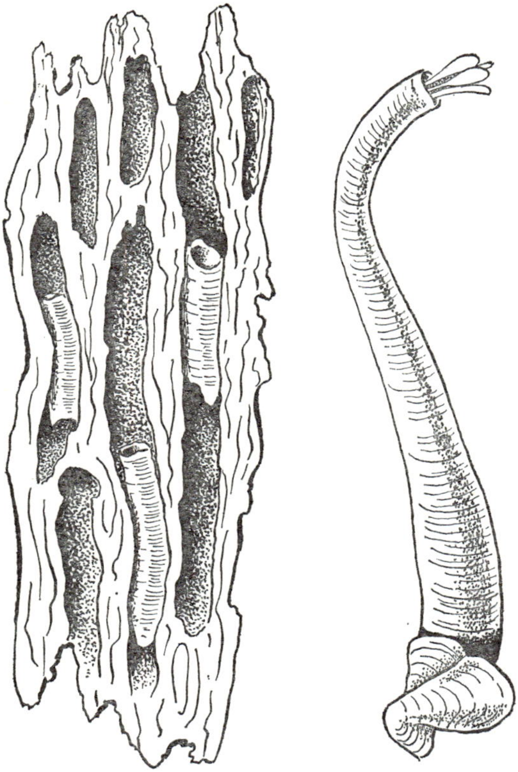

Fig. 67 Shipworm *Teredo norvegica*

find a large number of various-sized single valves or complete empty shells of other Bivalve Molluscs lying cast up on the shore. Some of these are frail but extremely beautiful in coloration, patterns, or shape, and are frequently collected perhaps for use as ornamentation on trinket boxes, souvenirs, etc. The Bivalve Molluscs themselves are either shallow burrowers (Artemis, Venus, Carpet, Wedge and Sunset Shells) or deep burrowers (Gaper, Otter, Tellin and Furrow Shells) in sand, gravel or muddy sand, depending on the species, of the lower shore and in deeper water and to be seen would usually have to be dug for in these habitats.

ARTEMIS, VENUS AND CARPET SHELLS [*Veneridae*]

BANDED CARPET SHELL (*Venerupis rhomboides*) Fig. 68 (a)

Common. It frequents deep water where it occurs in shell debris, burrowing in gravel, etc. The empty valves are very often found washed ashore. The oblong, glossy shell has flattened ridges. It is pale cream or greenish-white ground with brownish-purplish markings in an irregular, zig-zag fashion and measures up to 4 inches in length.

BANDED VENUS (*Venus fasciata*) Fig. 68 (b)
Very common. It frequents the middle and lower shore where it occurs in sand and gravel. The solid, not frail, triangular shell has obvious concentric ridges. It is very variable in colour, various yellows, browns, pinks, reds, white, as the ground, with darker rays, streaks or blotches running from the umbo or raised protuberance to the rounded edge of the shell. It measures up to 1¼ inches.

CROSS-CUT CARPET SHELL (*Venerupis decussata*) Fig. 68 (c)
Locally common. It frequents the lower shore, occurring in mixed, stony, muddy sand, sand, and muddy gravel, sometimes where the shore is rocky. Similar to the Pullet Carpet Shell but more square in shape, the shell has many prominent, coarse ridges which, where they meet the radiating "ribs", create a chiselled or engraved appearance, hence its name. The shell colour is usually a dull yellow with dark brown or purplish markings. It measures up to 2 inches in length.

OVAL VENUS (*Venus ovata*)
Common. It frequents the lower shore and deeper water where it occurs in sand. The shell has a cockle-like appearance, rather oval, with up to 50 "ribs" radiating from the raised protuberance or umbo on one edge, crossed by a series of concentric narrow grooves. The shell colour is yellowish or brownish-red with a pink or red variegation. It measures up to ¾ inch.

PULLET CARPET SHELL (*Venerupis pullastra*) Fig. 68 (d)
Very common. It frequents the lower shore and deeper water, where it occurs in muddy gravel, among other empty shells or shell debris and stones, at the foot of rocks. The oval shell is similar to that of the Cross-Cut Carpet Shell but the radiating "ribs" and ridges where they meet are not so prominent so the shell has a smoother, less chiselled appearance. The shell colour is variable, the ground is often a yellowish tint, sometimes with brownish irregular, variegated areas on it. It has a supposed resemblance to the plumage of a pullet, hence its name. The shell measures up to 2½ inches.

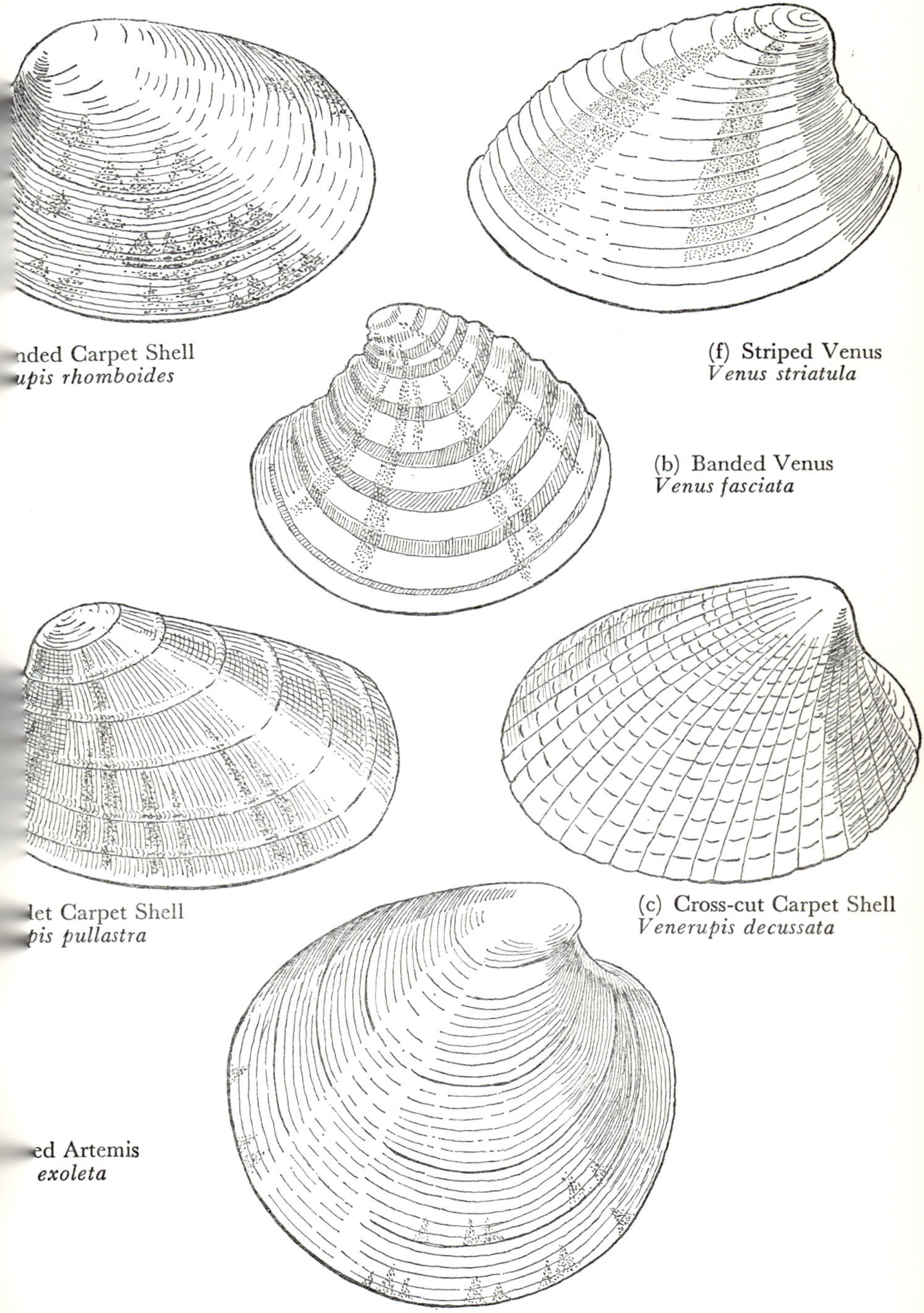

Fig. 68 Various Bivalve Shells

RAYED ARTEMIS (*Dosinia exoleta*) Fig. 68 (e)
Common. It frequents the lower shore where it usually occurs in sand, particularly in bays. The shell is nearly circular with smooth ridges. It is pale yellowish, greyish or brownish with several even paler ray markings from the raised protuberance to the edge of the shell. It measures up to 2 inches in diameter.

STRIPED VENUS (*Venus striatula*) Fig. 68 (f)
Common. It frequents the lower shore where it occurs in sand. The triangular shell has numerous overlapping concentric ridges. It is pale yellow, buff, light brown, with three ray markings, from the raised protuberance to the edge, reddish-brown in colour. It measures up to $1\frac{1}{4}$ inches in length.

WARTY VENUS (*Venus verrucosa*)
Common. It frequents offshore deeper water where it occurs in shell debris and gravel. The empty shells are sometimes washed ashore. It is easily identified from other Venus species by the concentric ridges which form rounded protuberances or tubercles on the thick shell, which have a wart-like appearance, hence its name. The shell colour is dull brownish-yellow and it measures up to $1\frac{1}{2}$ inches in length.

WEDGE SHELLS AND SUNSET SHELLS (*Donacidae and Asaphidae*)
BANDED WEDGE SHELL OR PURPLE TOOTHED SHELL (*Donax vittatus*) Fig. 69 (a)
Very common. It frequents the middle and lower shore where these are exposed and composed of clean sand, such as in bays. This is one of the most beautiful bivalve shells to be found while beachcombing. It is elongated in shape and compressed, with a polished, very glossy surface. The exterior of the shell is brightly coloured and varies being either white, yellow, brown, olive, bluish, purple or violet, sometimes with white ray patterning, but the inside of the shell is often violet. It measures up to $1\frac{1}{4}$ inches in length.

FAROE SUNSET SHELL (*Gari fervensis*) Fig. 69 (b)
Common. It frequents the lower shore where it occurs in sand. The shell is oblong elongated in shape and compressed, with concentric ridges. Its exterior is pink with rays of a different colour, sometimes white or yellow, radiating around the shell, while the shiny inside is usually purple. It measures up to 2 inches in length.

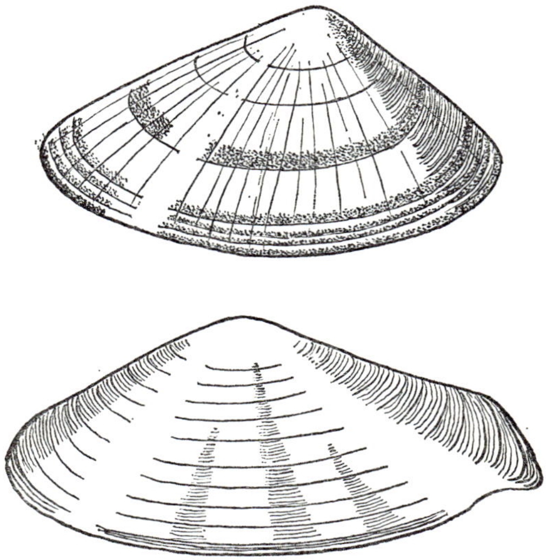

Fig. 69 (a) Banded Wedge Shell *Donax vittatus*
(b) Faroe Sunset Shell *Gari fervensis*

THE GAPERS (*Myidae*)

These are so-named because the large oval shells gape widely even when the bivalve shell is "closed" by the mollusc inside. They burrow vertically deep into the habitat, where they remain most of their lives, feeding on the plankton by means of their long siphons and have to be dug out to be seen, but the empty shells are cast ashore.

BLUNT GAPER (*Mya truncata*) Fig. 70 (a)

Common. It frequents the lower shore where it occurs buried in sand or a mixture of gravel and stiff mud. The shell has numerous irregular grooves. One end of the shell is rounded while the other is shortened, blunted and almost straight as if damaged, hence its name. It is brownish in colour and measures up to 2½ inches in length.

SAND GAPER (*Mya arenaria*) Fig. 70 (b)

Very common. It frequents the lower shore and commonly occurs in estuaries, in firm sand, mud or muddy sand. This Gaper is edible and also occurs on the United States coasts, where it is caught for consumption in large quantities and known as the "American Soft Shelled Clam". The shell is oval in shape, brownish or greyish in colour and measures up to 5 inches in length.

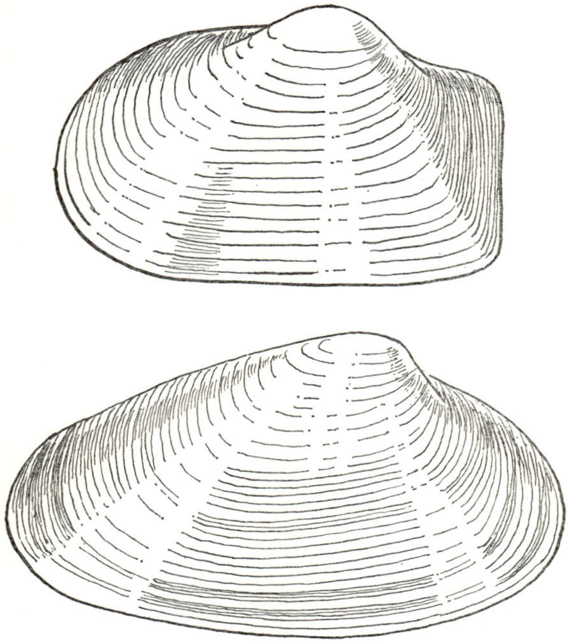

Fig. 70 (a) Blunt Gaper *Mya truncata*
(b) Sand Gaper *Mya arenaria*

TELLIN AND FURROW SHELLS (*Tellinidae and Semelidae*)
BALTIC TELLIN (*Macoma balthica*)

Very common. It frequents the lower shore, particularly being common in estuaries, where it occurs in sand, mud and gravel. The very flattened grooved shell is rounded in shape. It is dull red and pure white in colour, but many other colour variations, from pale yellow to dark red, sometimes with banded patterning, are to be found. A deep sea variation is orange-red and white. The shell measures up to 1 inch in length.

PEPPERY OR FLAT FURROW SHELL (*Scrobicularia plana*)
Fig. 71 (a)

Common. It frequents the exposed lower shore, but also occurs in estuaries, mudflats and salt marsh inlets and close to regions where the fresh water of rivers joins the tidal waters of the estuary. It buries itself deep in soft mud and muddy sand. The thin, very flattened shell is rounded and dull yellowish, brownish or greyish in colour. It measures up to 2 inches.

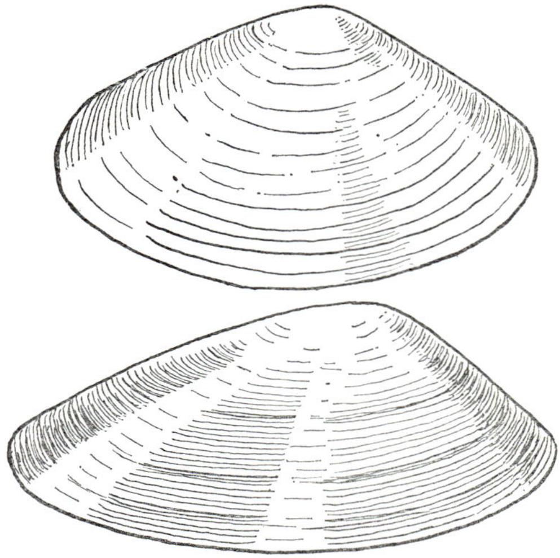

Fig. 71 (a) Peppery or Flat Furrow Shell *Scrobicularia plana*
(b) Thin Tellin *Tellina tenuis*

TELLINA FABULA

Very common. It frequents the lower shore where it occurs in sand. The very flat, polished shell is oval, with delicate tinted coloration of pearl-white, cream, pale yellow, pink, pale blue, sea green and other colours. It measures up to $\frac{3}{4}$ inch.

THIN TELLIN (*Tellina tenuis*) Fig. 71 (b)

Very common. It frequents the middle and lower shore, where it occurs in sand. The thin, very flat, polished shell is oval, with delicate and beautiful tinted coloration of deep pink, yellow, orange, white, cream, pale blue and other colours. It measures up to $\frac{3}{4}$ inch.

Chapter 9

Molluscs – Various

There are a considerable number of Molluscs, the *Opisthobranchia* Gastropods, which have little outward resemblance to the Univalve or Bivalve Molluscs because their external shell is either so minute that the animal is not able to withdraw into it, the shell is very small and enclosed within the Mollusc's body, or the shell is absent completely. For this reason many of these Molluscs have a superficial resemblance to the land slugs, but their bodies are similarly constructed, with a "foot", "mantle", "head", "mouth" and "radula", "gill" and "gill cavity". Even so, most of them blend remarkably with their background or prey and can be difficult to detect until removed from this situation.

The following, *Tectibranch Sea Slugs,* have a small external or internal shell—

BERTHELLA PLUMULA
Common. It frequents the lower shore where it is found on and under rocks and stones, also in rock pools. It feeds on sea squirts and is more common in the habitats where its prey is likely to be found. The shell is completely enclosed in the body. It has a soft, yellow body with whitish markings and measures up to 1½ inches in length.

ROUGH SIDE-GILLED SEA SLUG (*Pleurobranchus membranaceus*)
Locally common. It frequents the lower shore during summer when it arrives to prey on sea squirts. This it achieves by crawling alongside or upon a sea squirt and forcing its proboscis down one of the sea squirt's siphons to feed upon the internal tissues of the victim. The shell is internally enclosed. Its soft, oval, flattish body has a brown-red mantle or back and a wide, darker-reddish wavy-edged foot. To deter enemies

its body is able to emit a distasteful acid noxious to fish. It measures up to 5 inches.

SEA HARE (*Aplysia punctata*) Fig. 72
Common. It normally frequents deeper waters where it occurs on oar weed and sea lettuce, but during the early summer it moves inshore to spawn, the numerous eggs being laid to form long orange or salmon-pink "strings" round seaweeds, rocks and sometimes other objects. It is also able to swim with two lobes on its foot, and has the means if molested of emitting a purple liquid from its mantle, to deter being handled or siezed. It also contracts its body into a shape similar to the Beadlet Anemone and thus protects it from being interfered with by fish. The semi-transparent, thin, flat, oblong shell is almost enclosed internally in the soft body. On its small head it has two pairs of tentacles, the upper pair, when raised, being long and having a supposed similarity to the ears of a hare, hence its name. The body is olive-green, yellowish-brown, brown, dark purple, red-brown or reddish, with white or other colour spots, the colour depending on its age and food plant which it blends with. It measures up to 6 inches.

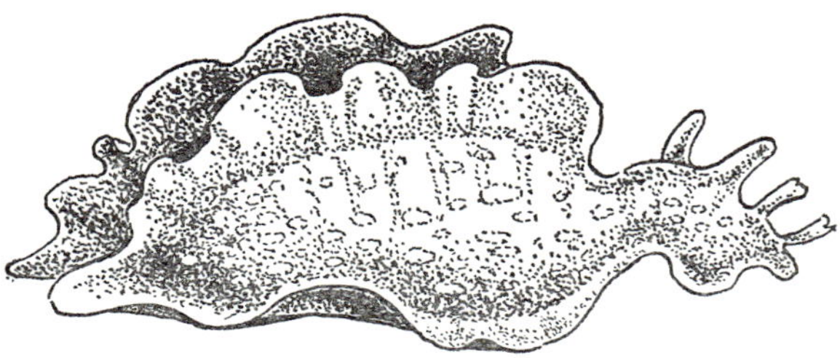

Fig. 72 Sea Hare *Aplysia punctata*

SOFT BUBBLE SHELL (*Akera bullata*) Fig. 73
Locally common. It frequents the lower shore where there is eel-grass, also occurring on mudflats. The thin, oval shell is external, but partly enclosed within the Mollusc's long, fleshy, semi-transparent, greyish body when feeding or dormant. It has an aperture the entire length of the shell but cannot withdraw into it. When it swims, by moving the two, large, flap-like foot lobes or extensions, the shell is suspended beneath it. This is one of several species of Bubble Shell which derive their name from their thin, inflated supposedly bubble-like appearance, but not all are rounded in shape.

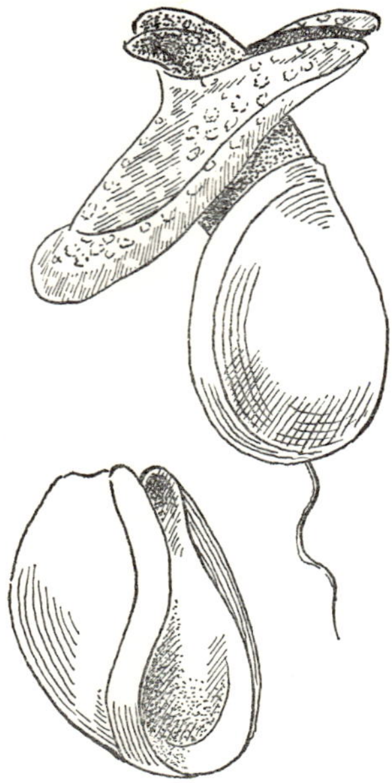

Fig. 73 Soft Bubble Shell *Akera bullata*

The following Molluscs do not have an external or internal shell. They are called *Nudibranch Sea Slugs,* which means they have exposed, naked gills which may look like feathery plumes, tiny anemones, or other extensions upon their body. Some of these Molluscs, despite their name, are beautiful in shape and movement. All of them are carnivorous. They also all lay eggs in a jelly-like material, the shape and colour of this depending on the species. From this the larvae hatch and live in the plankton until they return to the sea bed and develop into adults.

ANGLED DORIS (*Goniodoris nodosa*) Fig. 74 (a)
Common. It frequents the lower shore under rocks and stones where it feeds on sea squirts. The long, smooth body has a ridge or keel down its back. It varies in colour, may be whitish, yellow or rose-pinkish, with white or yellow spots or sometimes entirely white, with yellow tentacles. It measures up to 1 inch in length.

COMMON GREY SEA SLUG (*Aeolidia papillosa*) Fig. 74 (b)
Very common. It frequents the lower shore and is found upon rocks, stones or seaweeds, such as the wracks, where it feeds on sea anemones

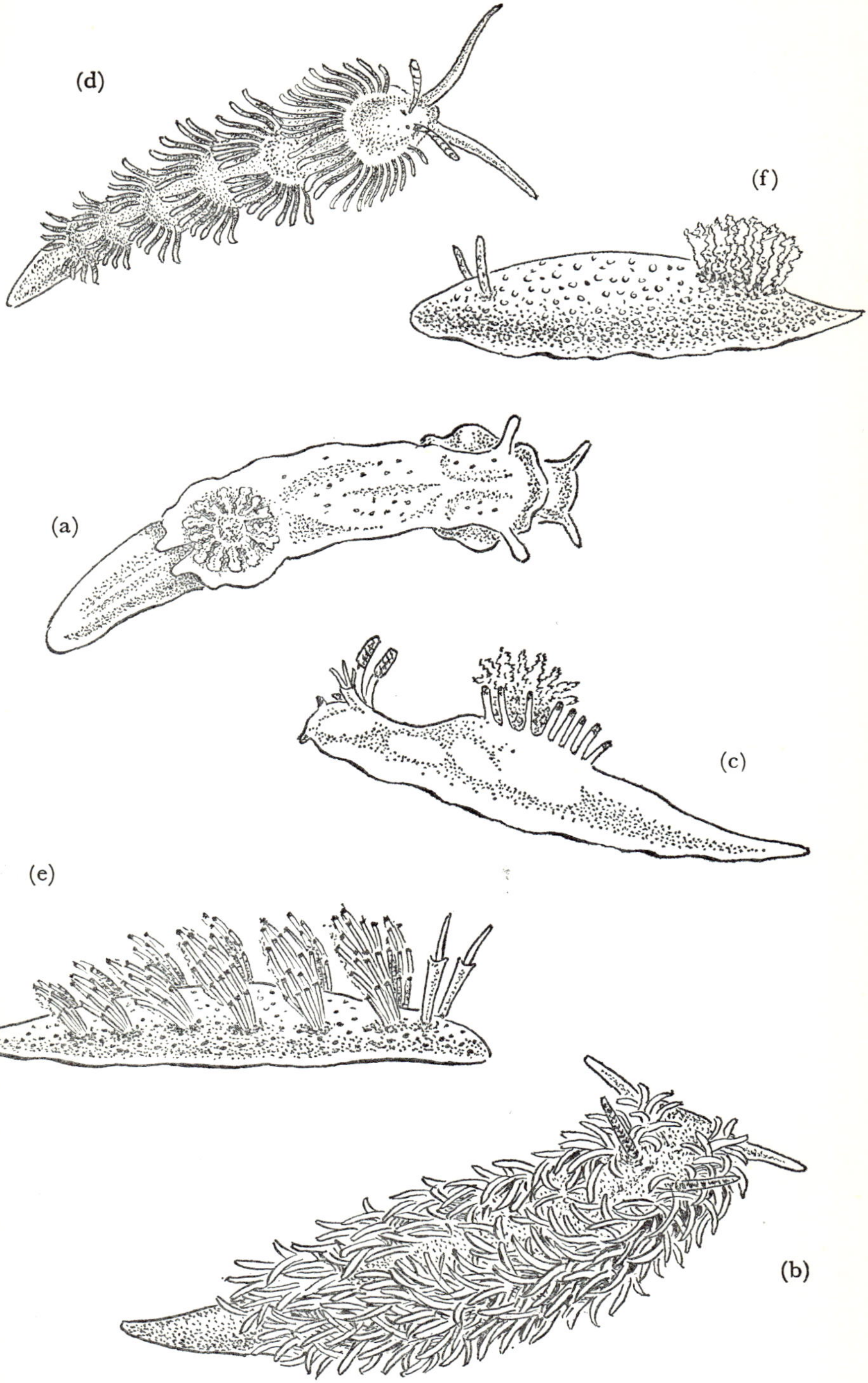

Fig. 74 Some Sea Slugs

) Crowned Eolis *Facelina auriculata*
) Angled Doris *Goniodoris nodosa*
Crowned Sea Nymph *Doto coronata*

(f) *Onchidoris muricata*
(c) Crested Slug *Ancula Eristata*
(b) Common Grey Sea Slug *Aeolidia papillosa*

and is unharmed by the latter's tentacles and stinging cells. This Sea Slug has the clever habit of placing some of the tentacles of a previously killed anemone on its back projections for protection from its own enemies. For this reason it is sometimes difficult to detect when it is feeding upon a similar anemone. The body, apart from down its centre, has numerous long, soft, brownish-greyish, greenish-brownish or pale buffish-brown projections, which give it the resemblance of wearing a shaggy fur coat. It is the largest British Nudibranch Sea Slug, measuring up to 3 inches in length, sometimes longer.

CRESTED SLUG (*Ancula cristata*) Fig. 74 (c)
Common. It frequents the lower shore where it is found among rocks and stones. The long body has tentacles around the gills resembling a tiny anemone on its back. In colour it is pale whitish, the head, tentacles, other projections and foot being tipped with deep yellow or bright orange. It measures up to 1 inch.

CROWNED EOLIS (*Facelina auriculata*) Fig. 74 (d)
Common. It frequents the lower shore, on rocks and stones where it preys on sea anemones. The long, narrow, body has up to six pairs of projections along its length. In shape it is not unlike a smaller, less "clothed" Grey Sea Slug to which it is closely related. It is whitish or yellowish, the projections being crimson-red, except for the tips which are white. It measures up to 1 inch.

CROWNED SEA NYMPH (*Doto coronata*) Fig. 74 (e)
Locally common. It frequents the lower shore on the undersides of rock and stones where it feeds on sea firs. The slender body has several pairs of narrow lobes or projections on each side, somewhat resembling pine cones. In colour it is pale whitish dotted with crimson, the projections also being similarly brightly marked. It measures up to ½ inch.

ONCHIDORIS MURICATA Fig. 74 (f)
Common, particularly on northern coasts. It frequents the middle and lower shore where it feeds on the sea mat (*Membranipora membranacea*). The oval, flattish-shaped body has pimple-like tubercles on its back. In colour it is yellowish with a brown or red spot or spots. It measures up to ½ inch.

RED DORIS (*Rostanga rufescens*)
Fairly common, more so on southern coasts. It frequents the lower shor

where it feeds on sponges, usually the Blood-Red Sponge (*Hymeniacidon sanguinea*). The body is elliptical and has small tubercles on its back. It is scarlet with a few black spots. Because of this colouring it matches the sponge it preys on and it is not easy to see thereon. It measures up to ½ inch.

SEA LEMON (*Archidoris pseudoargus*) Fig. 75
Very common. It frequents the lower shore where it is found on rocks, stones and seaweeds in pools, feeding on encrusting sponges, usually the Breadcrumb Sponge (*Halichondria panicea*). The white coils of up to 50,000 eggs, like ribbon, are commonly found in spring and early summer around seaweed, stones and small rocks. The body is elliptical in shape. Its colour is pale yellowish with brown, green, purple or pink markings, while the foot is yellowish, hence its name, but when it feeds on the blood-red sponge it may vary in colour to dull orange or reddish. It measures up to 3 inches.

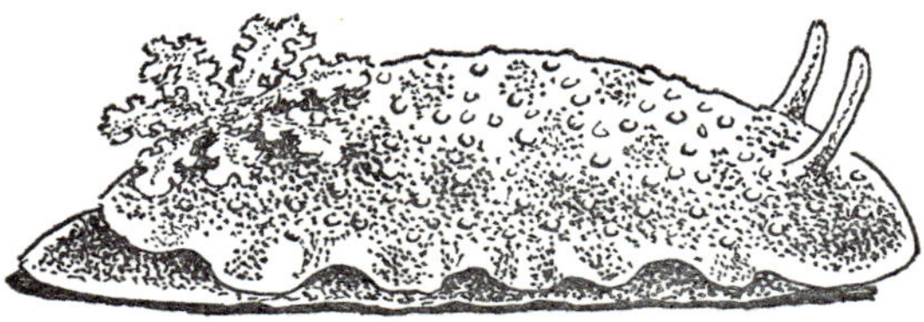

Fig. 75 Sea Lemon *Archidoris pseudoargus*

THE CEPHALOPODS

Because of their awesome, fear-inspiring reputation it may be surprising to learn that species of Octopus, Squid and Cuttlefish occur around the British coasts, also in estuaries, creeks and on mudflats. Fortunately none are as large or dangerous as species living in the oceans or the imaginative creations of fiction writers! It is also difficult to realise that these animals are related to the Univalve and Bivalve Molluscs but the Squids and Cuttlefish do have a rudimentary internal shell, although in the Octopus it is only a very small vestige. They belong to the *Cephalopoda,* from the Greek meaning "head-footed", referring to their anatomy in which the head and foot are fused together and extended to form the tentacles which surround the head. These are used to capture and hold the prey, as all these species are carnivorous or flesh-eaters. On the tentacles are rows of vacuum suckers which do the seizing, by gripping a victim powerfully. Inside the mouth they possess strong jaws likened to a parrot's beak upside-down and within the jaws the radula or "teeth". The Cephalopods have two, large, prominent, unwinking eyes, beneath the tentacles, with lenses, providing them with keen eyesight for capturing food or themselves avoiding attack. The gills are contained within the muscular mantle and cavity. By muscular exertion there is an intake of water for respiration and similarly by contraction this water is expelled in a jet from a funnel or siphon on the front underside of the body. This jet force from the funnel, whose direction can be controlled, is used by the animal to move swiftly in the water. Close to the funnel and connected to it there is the "ink-sac" containing the sepia or black fluid. If the Cephalopod is threatened it ejects the "ink", via the funnel, into the water. This quickly forms a cloud surrounding the Cephalopod, confusing the enemy and allowing the Cephalopod to escape. These molluscs have a highly sensitive nervous system which controls the pigment spots in their skin and when they require to avoid detection they can immediately alter the pigment to blend with their background.

The *Cephalopoda* is divided into the Cuttlefish and Squids (*Decapoda*) and the Octopus (*Octopoda*).

The Cuttlefish and Squids have five pairs of tentacles. One pair of these are longer than the other four short pairs. They have suckers which assist in seizing a victim – fish, crab, etc. – and can be retracted if not being used. Squids are caught by sea fishermen as bait and may be found dead on the shore. Sometimes the Squid's horny, "beak-like" jaws or its slender, keeled, horny "pen", the length of its back, can be found on the shore, the latter so-called because it has a supposed resemblance to a quill pen. Cuttlefish may be caught in shallow water, but are more frequently found dead on the shore, with only the white, porous "cuttlebone" remaining. The eggs of the Cuttlefish, normally attached to large seaweeds, are also familiar objects washed up on the shore during the spawning season, being purplish-black and rather like a bunch of small grapes.

The Octopus have eight tentacles, each the same length, armed with suckers. Although it can swim when required the Octopus, with its rounded, bag-like body and long tentacles, is more suited for crawling over the seabed. It lurks among rocks lying in wait for victims, but can be found hiding under rocks and stones on the lower shore at low tide or similarly in rock pools.

COMMON CUTTLEFISH (*Sepia officinalis*) Fig. 76
Common. It normally frequents deeper water, but moves to the lower shore in shallower water of sandy bays and estuaries to lay its eggs in

mid-summer. Dead examples are more likely to be found, also the large, white, soft, chalky, lance-shaped, internal "shell" or "cuttlebone" which is commonly cast up on the shore. It might be thought the Cuttlefish has a high mortality rate when close inshore, but the usual reason for the presence of such numerous "cuttlebones" is they are preyed on by conger eels and large fish who relish cuttle as a fleshy morsel. The body is broad and has lateral fins, while the head is narrower than the body, with black eyes. It has black and white or brown and white stripes on the back like a zebra. If it is chased or attacked it has the ability to quickly change its colour. The "cuttlebone" of this species is the familiar object bought by cage bird lovers for their pets to gnaw at with their beaks. The body, excluding the tentacles, measures up to 12 inches in length.

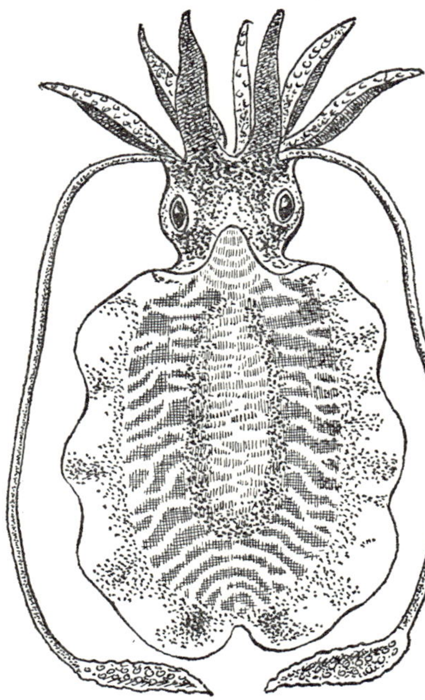

Fig. 76 Common Cuttlefish *Sepia officinalis*

COMMON OCTOPUS (*Octopus vulgaris*) Fig. 77
Common, but occurring mainly in the English Channel and thus on the south coast. It frequents the lower shore and can be found among rocks or stranded in pools. It has two rows of suckers on each tentacle

and feeds, using the horny "beak", mostly on bivalves and crustaceans. It varies in colour from mottled dusky brown to greyish, yellowish, purplish, spotted with brown, purple or orange. Considering its reputation it is rather disappointing to find that its body length is up to 6 inches, sometimes slightly more, the tentacles each being up to 24 inches in length.

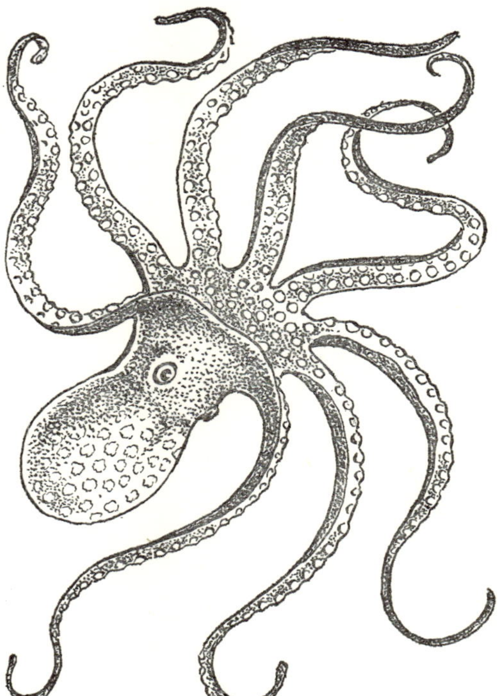

Fig. 77 Common Octopus *Octopus vulgaris*

COMMON SQUID (*Loligo forbesii*)
Uncommon alive on the shore, common in deeper water, or on th sea's surface, sometimes in shoals, but the occasional dead specimen ma be found cast up. More frequent are the beak-like jaws and skeleto "pen". The slim, cylindrical, blunt-pointed body has been likened t torpedo-shape, with two triangular lateral or side fins. This streamline body assists it to swim swiftly. Fortunately this Squid does not reac the gigantic proportions of the fictional deep-sea monsters and on findir one its body length of up to 24 inches is rather disappointing. Tl majority of examples found inshore are about 12 inches in length.

CURLED OCTOPUS (*Eledone cirrhosa*)
Common, but occurs more frequently on northern coasts, where it is found on the lower shore and in pools. It has one row of suckers on each tentacle. In colour the upperside is red-brown while the underside is a bluish-white with larger spots. The body length is up to 6 inches, the tentacles each being about 12 inches.

LITTLE CUTTLE (*Sepiola atlantica*)
Common. It frequents the lower shore, more so in sandy bays and is often captured in shrimping nets. It also burrows in the sand by using the jet of water from its funnel to create a hole for its body. The smooth body is rounded with two rounded, lobe-like side fins. It is flesh-coloured and measures up to 2 inches, with the tentacles about twice the length of the body or less, this being the smallest British Cuttlefish.

Chapter 10

Marine Worms

The very many Marine Worms occuring on the shore are divided into two groups, the Marine Bristle Worms (*Polychaeta*) and the Ribbon Worms (*Nemertini*), the latter having been described separately in this chapter to avoid confusion.

The Marine Bristle Worms vary considerably in anatomy, way of life and size, and there are more of them than the simpler Ribbon Worms. Some examples of *Polychaeta* are less than an inch in length, but others measure more than a foot. Their body is segmented, which means, unlike the Ribbon Worms, that it comprises small divisions or compartments, the segments having their own internal and external organs. They bear bristles, except the Tube-worms, some species having projecting, paddle-like lobes. Their food, depending on species, is either planktonic and other prey or organic matter. Reproduction is by eggs which hatch as plankton-living larvae until maturing and settling in the habitat.

The Marine Bristle Worms are also divided again into two categories, into either of which they fit according to their mode of existence. One group are called Errant Marine Bristle Worms because they have an active or semi-active crawling and swimming life. The second group are called Sedentary Marine Bristle Worms because they live their confined, sluggish lives within a burrow or specially-constructed tube. Space unfortunately does not allow the description of all the very large number of Marine Worms, but those most likely to be discovered and identified are included.

RAGWORMS—Errant

The larger Ragworms (*Nereidae*) are frequently gathered for bait. They have numerous segments with projecting bristles, which are used for moving along. The head is large with two antennae, several feelers

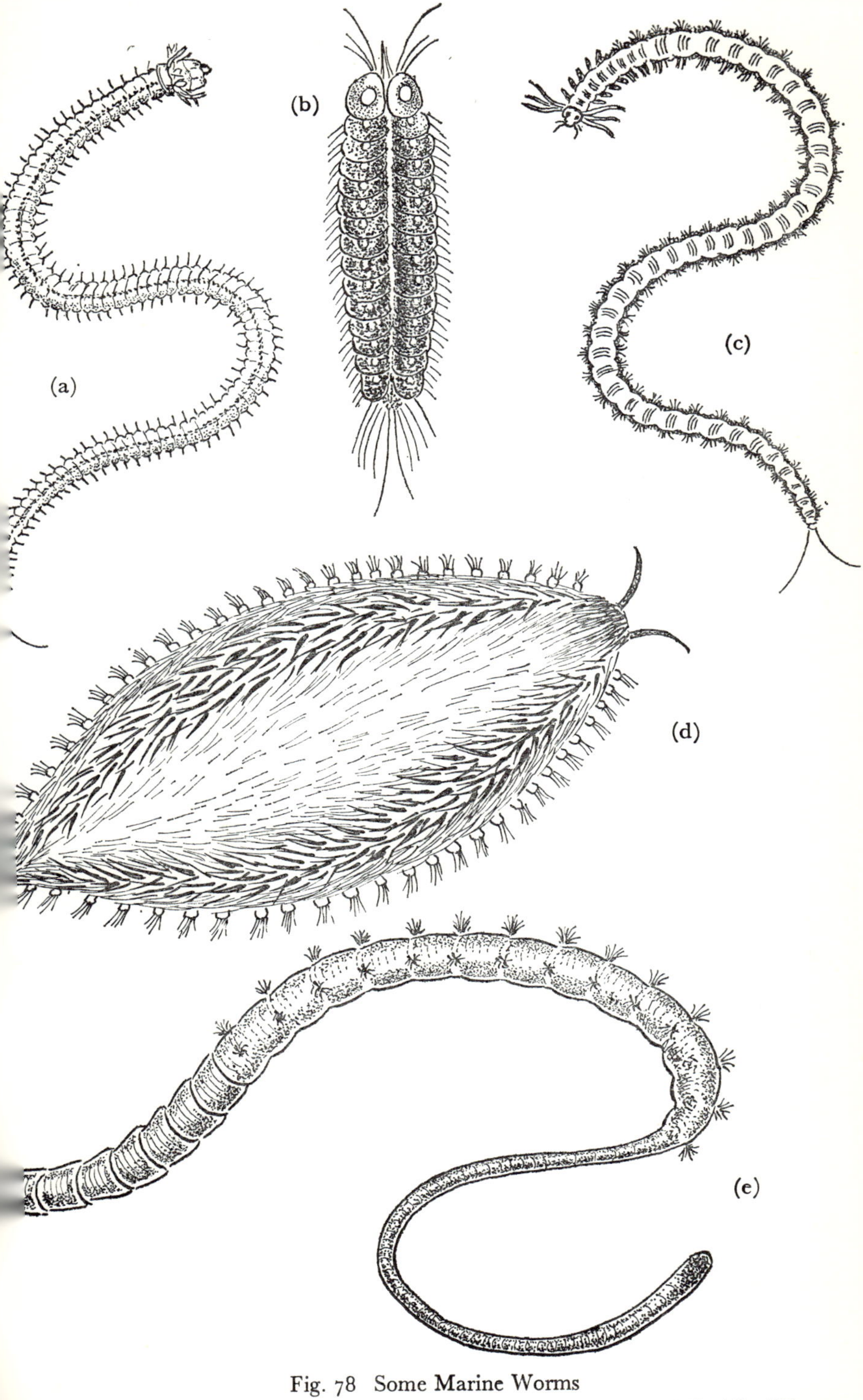

Fig. 78 Some Marine Worms

) *Nereis diversicolor* (b) *Harmothoë impar* (c) Spotted Paddle Worm *Phyllodoce maculata*

) Sea Mouse *Aphrodite aculeata* (e) Lugworm *Arenicola marina*

and four eyes. They also have a pair of powerful black jaws at the end of a short proboscis. One species, *Nereis fucata*, has the habit of living within a whelk shell while it is also occupied by the hermit crab, both existing in harmony and gaining mutual benefit from the other's presence.

NEREIS DIVERSICOLOR Fig. 78 (a)
Very common. It frequents the middle and lower shore, where it can be found in or on sand and mud and also occurs in river estuaries, mudflats and brackish water areas. It is variable in colour, but usually an orange-brown or yellow-brown with a greenish shading on the sides, while a conspicuous red line, a blood vessel, runs the full length of the back. It has over 100 bristled segments and measures up to 4 inches in length.

NEREIS VIRENS
Locally common. It frequents the lower shore where it can be found under stones, rocks, in muddy sand. This Ragworm when full grown is almost unmistakable due to its size and the finder should curb a desire to pick it up recklessly, as care is needed to do so, or it will inflict a painful bite on the skin with its jaws. It is green in colour with purplish irridescence on the back, while the underside has a pale pink tint. Usually it is about 12 inches in length, although examples up to 36 inches and as thick as a middle finger have been found.

PERINEREIS CULTRIFERA
Common. It frequents the lower shore where it occurs in sand, gravelly mud or under stones, also in pools. It is grey-green or bronze-green in colour. It has over 100 bristled segments and measures up to 10 inches in length.

SCALE WORMS—Errant
The Scale Worms (*Aphroditidae*) are so-called because the back of their somewhat flat bodies are covered entirely or partly by pairs of protecting, hard, interlocking scales. If seized they discard their scales which may be an attempt to confuse an enemy so that they can escape although some species disintegrate if touched. One species, *Scalisetosus assimilis*, which measures up to an inch in length and is pale with transparent scales, except for a dark line on length of the back, lives only among the spines of the common sea urchin (*Echinus esculentus*).

HARMOTHOË IMPAR Fig. 78 (b)
Common. It frequents the middle and lower shore where it occurs under seaweed and stones. In appearance it has a similarity to the land centipede, with 15 pairs of overlapping, roundish or kidney-shaped hair-fringed scales covering the entire body. The colour is brown or greyish-brown, with a yellow mark near the centre of the scales. It measures up to an inch in length.

SAND SCALE WORM (*Sthenelais boa*)
Locally common. It frequents the lower shore where it occurs under stones or in sand. It is long and thin with about a 150 pairs of overlapping scales which it does not discard if handled. It varies in colour, most often greyish-white, and measures up to 8 inches.

SEA MOUSE (*Aphrodite aculeata*) Fig. 78 (d)
Common. It usually frequents deeper water offshore, but sometimes can be found on the lower shore, on or in sand or muddy sand, scavenging on carrion. An aid to identification from other species is that its scales are obscured by a thick protective covering mat of fine hairs and thin bristles, hence its name. It is a flat, long, broadly oval shape, with 15 pairs of large scales. The colour is greyish on the back, the sides having brilliant irridescent green and gold hairs and red-brown spines. Its appearance is misleading and not wormlike, being the largest British Scale Worm, measuring up to 6 inches in length.

PADDLE WORMS—Errant
The Paddle Worms (*Phyllodocidae*) are so-called because their body has projecting lobes shaped like paddles. They do not have body scales or jaws on their proboscis. The Paddle Worms swim agilely. The following are three that should be identified without difficulty.

GREEN LEAF WORM (*Eulalia viridis*)
Common. It frequents the middle and lower shore, where it occurs in the cracks in rocks and also upon them in pools. It is especially likely to be found in areas where barnacles, its prey, are abundant. The body is long and narrow, the head rounded, with five antennae. In colour it is a shiny, dark green, with pale green near the head and lighter green pointed paddles. It measures up to 3 inches in length.

GREEN PADDLE WORM (*Phyllodoce lamelligera*)
Uncommon. It frequents the lower shore, where it occurs under stones

and among laminaria brown seaweeds. In colour it is an irridescent blue-green on the back, with green paddles. The body has over 400 segments and measures up to 18 inches, sometimes longer.

SPOTTED PADDLE WORM (*Phyllodoce maculata*) Fig. 78 (c)
Fairly common. It frequents the lower shore where it occurs on and under rocks and stones where these are partly buried in sand. The head is heart-shaped, the body being more narrower than the Green Paddle Worm. In colour it is green or yellow, with brown dots or marks on the back segments, and brown paddles. It measures up to 4 inches.

LUGWORMS—Sedentary

When men are seen on the tidal flats at low water digging and from time to time dropping something into a bucket they are almost certainly gathering Lugworms or Lobworms (*Arenicolidae*) for use as bait. These worms live a sluggish life, each in a U-shaped burrow just below the surface of the sand or muddy sand. They use mucus from their body to strengthen the sides of the burrow and prevent it collapsing. Their presence is indicated by the conspicuous heaps of coiled sand, the worm "casts", with, a few inches away, a shallow pit or depression. The Lugworm feeds by sucking down and swallowing large amounts of sand at the site of the depression and small hole, digesting the edible organic matter in this, then getting rid of the unrequired sand and waste matter by forcing it out of the other open end of the U burrow, to form the coil or "cast". When the tide returns it washes the heap level and also fills in the depression caused by the Lugworm feeding, providing it with another supply of sand. The Lugworm population is highest where the sand is rippled on the surface, thus retaining some of the water and having a suitable liquidity. Three species occur in Britain; one is common, the others being less frequent.

ARENICOLA BRANCHIALIS
Fairly common, more so on south and west coasts. It frequents the lower shore, in fine gravel under stones. The body has about a dozen segments and red gills, with short bristles in front of the first two gills. In colour it is a greenish-black or yellowish-black and measures up to 8 inches.

ARENICOLA ECAUDATA
Fairly common. It frequents the lower shore where it occurs in fine gravel among stones. The body has about 16 segments, and red gills

and short bristles along its full length. In colour it is greenish-black or yellowish-black and measures up to 8 inches.

ARENICOLA MARINA Fig. 78 (e)
Very common. It frequents tidal flats, coastal sandy bays, estuary sand and muddy sand. The soft, plump body has shortened bristles with 13 pairs of red feathery gills in its middle region, while the more rigid "tail" or hind region is narrower and does not bear gills or bristles. It has a large proboscis which it thrusts outwards from the mouth to consume the sand and also when necessary grip the burrow wall to assist the bristles in the same purpose. In colour it is greenish-black, brownish-black or yellowish-black and measures up to 9 inches. This is the most well-known and familar Lugworm or Lobworm.

TUBE-LIVING WORMS—Sedentary

There are a large number of Sedentary Worms which can construct tubes, either on the surface of various seaweeds, protruding from the mud, sand or fine gravel or on rocks and stones. Some of the tubes are made of limy material, sand, mud, or small shell fragments and similar material bonded with a mucus from the Worm's body and are so delicate that they collapse on being touched. The Worms feed by using their tentacles to secure minute organisms from the water. Their means of escaping danger is to hopefully retreat quickly into their tube.

HONEYCOMB WORM (*Sabellaria alveolata*) Fig. 79 (a)
Common. It frequents the middle and lower shore in rocky, sandy places, also in suitable estuaries. As its name describes, colonies of this Worm fuse their numerous hard tubes together from sand grains so that they resemble honeycombed sandstone. These are created on semi-exposed rocks embedded in surrounding sand and the mass may measure up to 12 inches or more across. The Worm has short thread-like tentacles and a golden fringed "collar" around its head, projected when covered by the tide to gather sand or food organisms. In colour it is red or brownish-red and measures up to 1½ inches.

KEEL WORM (*Pomatoceros triqueter*)
Common. It frequents the lower shore where it occurs upon stones, rocks, shells and fixed, usually smooth, flattish-surface objects. Here it creates a long, twisting, dull white, limy tube, with a triangular cross-section and starts from a curly, very thin "tail" to a sharp point at the tube's opening. On the top runs a keel, hence its name. When covered

by the tide the Worm projects its crown of red or orange spotted and barred white gill tentacles, the main body being a purplish-brown. The tube measures up to 4 inches in length.

PEACOCK WORM (*Sabella pavonia*) Fig. 79 (b)
Common. It frequents the lower shore where it occurs in estuaries and on tidal flats of sand, mud or gravel. The long, flexible, upright, round tube is created from sand or fine mud particles combined with slime produced by the Worm. Although it may be up to 18 inches in length only some 4 or 5 inches of this, sometimes slightly curving, projects from the sand, mud or gravel during low tide, the hidden portion being attached to a buried stone or rock. When covered by the tide the Worm projects a wide crown or ring of brightly coloured gill tentacles, in reds, violets, browns, etc., with darker bands of colour, hence its name. These it uses for respiration and to gather food organisms from the water above the sand or mud surface, as well as for sand and mud particles for its tube. The Worm's body is a light green or greenish-grey and has tints of orange and violet near its tail. It measures up to 10 inches.

PECTINARIA KORENI Fig. 79 (c)
Common. It frequents the lower shore on and in sandy flats anc estuaries. The frail, bent, tapering tube is open at each end and createc from grains of sand cemented together. It is up to three inches in lengtl and as the Worm moves from one site to another it can convey its firn tube "home" with it. Its empty tubes are quite commonly found on th shore. The Worm has bristles on its 15 segments. In colour it is pinl or a brownish-pink and measures up to 2 inches.

POLYMNIA NEBULOSA
Common. It frequents the lower shore where it occurs on muddy, sand gravel, usually under rocks, stones, shells or similar objects. The fra tube is constructed of a viscid mucus slime with small shell fragment stones, gravel, attached to it, which in turn may be fixed to the unde side of the object. If the stone, etc. is lifted the tubes or portions of ther will be seen on its underside. The undamaged tube measures up to 1 inches in length. The soft, plump Worm has up to 100 segments, bu tapers towards the tail. Usually 17 of the segments have tiny warts an bristles. It has white marked red gills in three pairs behind the hea and projects a crown of white, yellowish or light orange threadlik tentacles from the head, the white-spotted orange body measuring u to 8 inches.

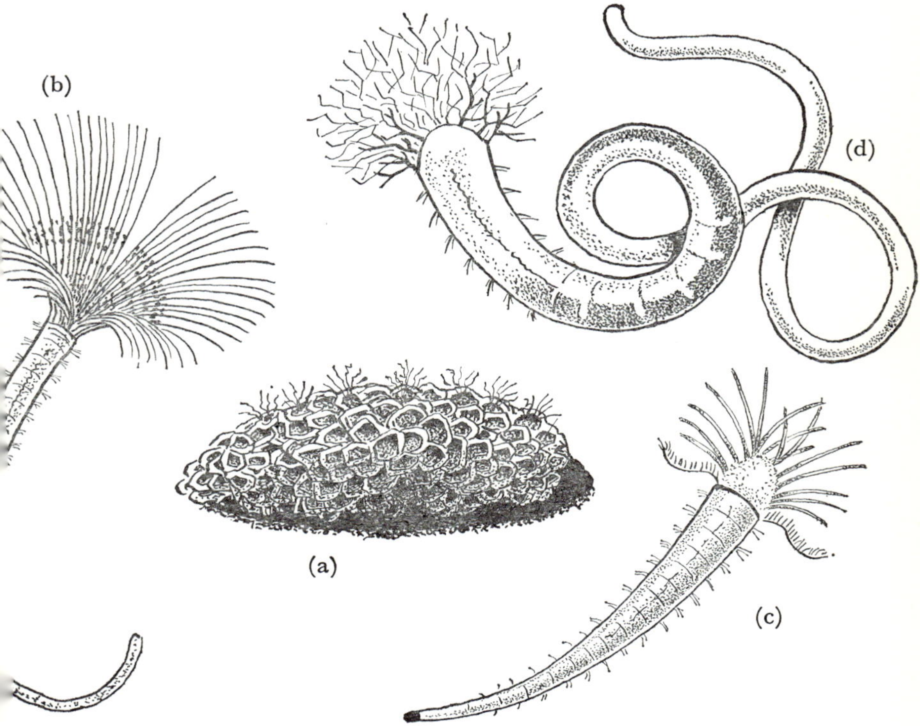

Fig. 79 Some Tube Living Worms
b) Peacock Worm *Sabella pavonina* (d) Sand Mason *Lanice conchilega*
(a) Honeycomb Worm
Sabellaria alveolata
(c) *Pectinaria koreni*

·AND MASON (*Lanice conchilega*) Fig. 79 (d)

'ery common. It frequents the lower shore where it is found in sand, specially where there are also some seaweed-covered rocks. The frail ıbe is constructed of a sticky mucus mixed with a coating of sand, nall gravel, stone and shell fragments. Although the tube is up to 2 inches in length only about two inches with a fringe of tentacles, sually projects upright from the sandy shore. It is a marvellous piece f delicate construction for such a creature but immediately collapses r breaks into pieces if attempts are made to pick it up. The tubes ary according to the site and on one shore where bivalve molluscs are ommon their fragments may dominate in the construction, whereas on sandier, gravelly shore these materials will dominate. The Worm has up to 300 segments, only 17 bearing the bristle tufts. The three pairs of gills are bright red and tentacles light pink. Its body is also pink, red or brownish and measures up to 10 inches.

SPIRORBIS BOREALIS Fig. 80

Very common. It frequents the middle and lower shore where it occurs, occasionally in vast numbers, on wracks, oarweeds and various other seaweeds, also cemented on rocks, stones and other objects. The tiny, hard, limy, white tube, like a flat snail shell, is coiled clockwise, approximately $\frac{1}{5}$ inch across. When uncovered by the receding tide the Worm retreats inside the tube but will project its greenish tentacles if it is put in a pool.

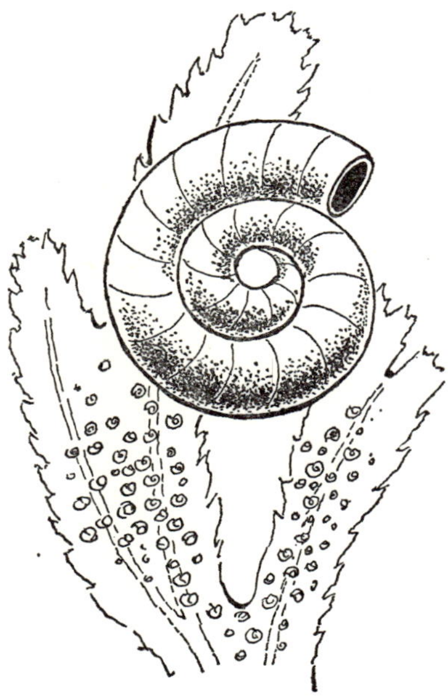

Fig. 80 *Spirorbis borealis*

RIBBON WORMS

The Ribbon Worms (*Nemertini*) are so-called because their body unsegmented, which means it does not comprise several divisions compartments. They are thus able to stretch their body, in son instances to extreme thin-ness and length or contract the body to short plumpness. Because of this it is only possible to give approxima lengths of these Worms and it is not unusual if some are thus foun much longer or shorter. Perhaps because of their extension powers it might have been more appropriate to name the group Cord or String Worms as Ribbon is not quite an accurate description. Their body is often slimy but smooth and progresses slowly, in a gliding movement, with wavy motions along its length from hind end to head. Some species have four eyes on their head, others have a varying number, while three species are eyeless. The Ribbon Worms are carnivorous. Their head has an opening over the mouth from which a long proboscis is suddenly pushed out to seize the prey. This proboscis sometimes has a needle-like instrument or may bear poison spines, but these would not harm a person handling these Worms. They reproduce by eggs which after hatching as tiny larvae live for a period in the plankton before descending to the habitat.

BOOTLACE WORM OR SEA SNAKE (*Lineus longissimus*) Fig. 81 (a)

Fairly common, more so on northern coasts. It frequents the lower shore where it occurs under rocks and stones, lying on muddy or sandy shingle, in a heap of intertwining coils. Sometimes it is also entangled in the oarweeds and holds on so thoroughly it cannot be unwound without damage. It is a blackish-brown colour with a purplish hue. Unmistakable because the slimy body is extremely long and it is the largest British Ribbon Worm. Examples up to five yards are not unusual. One has been recorded *30 yards* in length and at that stage it had only been half uncoiled: A giant's bootlace?

AMPLECTONEMA NEESII

Common. It frequents the middle and lower shore, where it occurs under small rocks and stones and when discovered it writhes itself into coils. The Worm has a flat underside and a rounded upper side. The upper side is brownish with streak markings and sometimes a purplish hue, the underside being whitish. Near the head, which has several eyes in clusters, there is a light streak across it like a dividing line. It measures 6 to 12 inches in length.

PINK RIBBON WORM (*Amphiporus lactifloreus*) Fig. 81 (b)

Common. It frequents the lower shore where it occurs under or among stones, shingle and seaweeds. The semi-transparent, flattened body narrows behind the head, giving the impression that the Worm has a "neck". It has four groups of eyes situated on the front and middle of the head. In colour it is a pinkish, greyish or whitish tint, with an

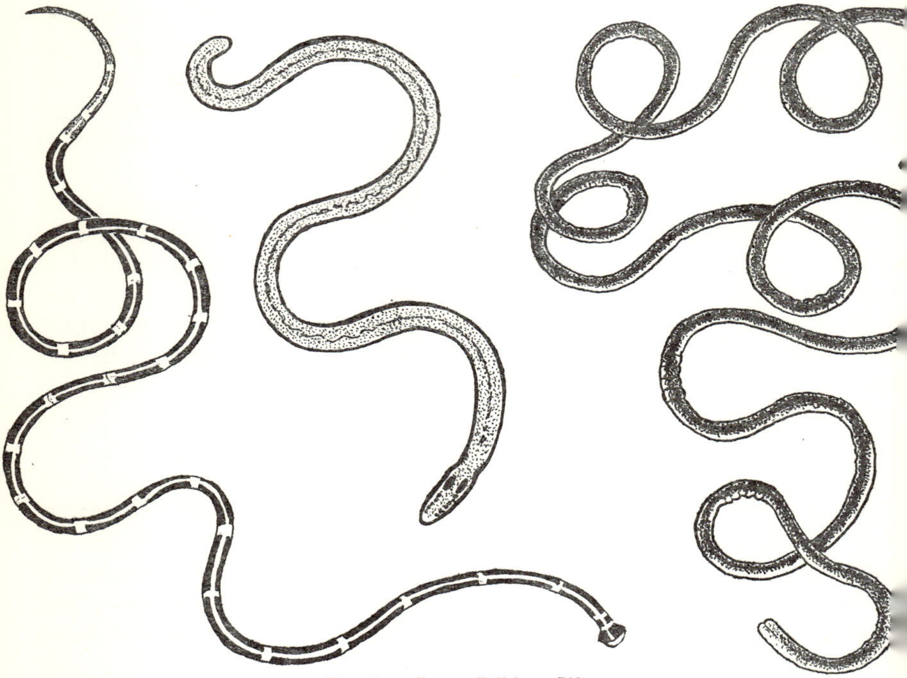

Fig. 81 Some Ribbon Worms

(a) Bootlace Worm *Lineus longissimus* (b) Pink Ribbon Worm *Amphiporus lactifloreus* (c) *Tubulanu annulatus*

irregular line on the back, also two pinkish-rose areas on its head. measures up to 3 inches.

TUBULANUS ANNULATUS Fig. 81 (c)
Fairly common. It frequents the lower shore where it occurs on san flats in a mucus tube, under stones and other debris. This is another the Tube Worms without eyes. It is rather beautiful, the thin bo being rusty-red or red-brown on the upper side, patterned with a wh striped "network", the underside being a lighter tint. It measures to 20 inches.

Chapter 11

Starfish and Relatives

STARFISH

The Starfishes (*Asteroidea*), with the Brittle Stars, Feather Stars, Sea Urchins and Sea Cucumbers, are members of the group *Echinodermata,* which means "spiny skinned". The body, strengthened by limy plates in the skin, is disc-shaped with five or more "arms" radiating from it, but these combine, so there is no clear division between the body disc and arms. There is no head, the mouth, which can be extended, being on the central underside of the body. The tiny red eyes are positioned one at the tip of each arm. Most remarkable is the internal series of tubes or canals forming the water vascular system. These connect with the sea water, by means of the perforated plate, he madreporite, on the body's upper surface, and take it throughout he body, including the rows of sucker-like, hollow, tube-feet under each arm, which can be swollen at will and which the starfish uses to grip and move. These also right the starfish if it is turned upside-down by the tide or other means. It pushes the sucker-bearing tube-feet forward in several directions until it is able to grasp an object and turn itself the correct way up. Sometimes in its efforts to do this the starfish bends one of its arms almost completely over until the tube-feet can reach something to get a grip on. When on a rock, slippery chalk foreshore or anything similar the tube-feet, by means of suction and mitting a sticky fluid, keep the starfish in position. Starfish frequently ose one arm or more in accidents but this is no impediment as they row a new one or several, which are sometimes shorter than the riginals.

The life cycle begins as an egg. When it hatches the young starfish oes not have arms or body disc. It lives as part of the plankton in the pper surface of the sea until it sinks to the sea bed and develops into n adult starfish.

The starfish feed on other marine creatures, including shellfish (molluscs) and when too common can wreak havoc in an oyster or cockle bed. Being very strong, the starfish uses the tube-feet on its arms to grip and force the two shells of the bivalves open. It has no teeth and usually cannot swallow the larger victims, for it has only a small mouth opening. So where it cannot force open the shell, such as that of a whelk or other univalve, it projects its stomach through its own mouth into the victim's shell and literally consumes the victim at home.

ASTROPECTEN IRREGULARIS Fig. 82

Common, but usually seen only if cast ashore in stormy weather. It burrows in sand offshore below low water. Varied in colour, reddish-violet, rust-red, pink, yellow, sometimes with purple tips on the arms. The body is very flat, the arms having a raised margin with a fringe of projecting spines. The five arms differ from other starfish in having pointed, not sucker-like, tube-feet which are adapted for rapid digging and almost useless for moving on rocks, etc. The suckers being absent, it cannot open shellfish so swallows these, also crustaceans, worms, fish, etc., whole, the shell, bones, etc., being disgorged after the flesh has been digested. It measures up to 5 inches in diameter.

Fig. 82 *Astropecten irregularis*

COMMON STARFISH OR FIVE-FINGERED JACK (*Asterias rubens*)
Very common. It frequents the lower shore, sometimes found in dozens at low water lying on mud, sand, among rocks and stones, or in pools. In colour it is usually brownish-yellow, orange-yellow or reddish-brown with paler markings; sometimes reddish-yellow, violet, or purple. It normally has five arms, but sometimes has six or four, tapered and rounded, rough to the touch, with a line of spines down the middle of each. Examples on shore usually measure up to 4 inches in diameter, but in deep water up to 20 inches.

COMMON SUNSTAR (*Solaster papposus*) Fig. 83 (a)
Fairly common. It frequents the lower shore, among rocks. The number of arms varies from 10 up to 12, occasionally 14. It also varies in colour, but the broader disc body is usually prominently red, purple-red, orange or yellow above, with arms paler, especially at the arm tips. Sometimes there is a banded pattern on the arms. Underside is whitish. The upper surface is rough with rows of spines. As well as the usual starfish diet it also preys on other starfish species. It measures up to 14 inches in width in deep water, but smaller examples usually found are up to 8 inches in width.

PURPLE SUNSTAR (*Solaster endeca*)
Not common. It frequents the lower shore, except on south-eastern and southern England coasts. It has more than five arms, the number varying between 7 up to 14, but usually 9 or 10. It varies in colour, being violet or purple, sometimes a reddish-yellow; the underside is whitish-cream. The upper surface is rough. It measures up to 16 inches in width but 6 to 10 inches is more usual.

SCARLET STARFISH (*Henricia sanguinolenta*)
Fairly common, particularly on the east coast of England. It frequents the lower shore, also in deep water. It has five arms, stiff, thin, tapering and rounded. Despite the name it is variable in colour and more usually crimson-red or purple-red, sometimes orange; pale orange-yellow on the underside. It measures up to 10 inches in diameter, more usually 5 or inches.

SPINY STARFISH (*Marthasterias glacialis*)
Fairly common, more so on the south and west coasts of England. It

Fig. 83 (a) Common Sunstar *Solaster papposus*
(b) Cushion Star or Starlet *Asterina gibbosa*

frequents the lower shore. It has five arms which it quickly discards picked up or trapped by one or more. The body is soft, rough to t touch, with knobs and spines, the latter particularly prominent in ro on the arms. The colour varies, pale yellow, orange, reddish, mau greyish-green or pale green. It measures up to 28 inches in diamet but is usually up to 12 inches.

THE CUSHION STARS

These species are also starfish, but described separately because of their shape or size. Their body disc is flat, with very short, blunt, reduced arms and no clear demarcation between disc and arms. They are immediately recognisable from starfish, which have a pronounced body disc and arms.

CUSHION STAR OR STARLET (*Asterina gibbosa*) Fig. 83 (b)

Common, but only on west and south coasts of England. It frequents the lower shore, on and under stones, on projecting rocks and similar. It has five short arms; the flat, stiff body disc is large in comparison with the arms. It is rough to the touch; varies in colour, brown, yellow, green, with different green, cream, or red tints. It is the smallest British starfish. It usually measures approximately 1 inch in width, but larger examples are occasionally found.

PIN-CUSHION STAR (*Porania pulvillus*)

Uncommon. It may be found occasionally washed ashore, but usually lives beyond the lower shore level, except on the east coast of England and Scotland, where it is rare. The thick body disc has five very short arms and a smooth, greasy texture. It is usually scarlet red above, sometimes with red markings, and white underside; occasionally yellowish-white above. It normally measures about 2 inches in width, but examples up to 4 inches have been found.

THE BRITTLE STARS

The Brittle Stars (*Ophiuroidea*) are so-called because they easily, quickly and deliberately dispose of one or more of their five arms to escape if they are accidentally trapped or seized by an enemy. The severing occurs near the junction of the arm and the disc. The body disc is usually well-defined and small, about an inch in diameter, in comparison with the long, thin arms, the arm length being up to five times the diameter of the disc body. They have no suckers on their tube-feet and instead move by sinuously writhing and jerking their arms. Brittle Stars, often mutilated and armless, are usually likely to be found washed ashore after stormy weather, among seaweeds and other marine flotsam. Some species are extremely difficult to identify, even when the finder has examples of each to study together, so only the most common and easiest to recognise are listed.

Fig. 84 Some Brittle Stars
(a) *Acronida brachiata*
(b) Daisy Brittle Star *Ophiopholis aculeata*

ACRONIDA BRACHIATA Fig. 84 (a)

Fairly common, on south and west coasts of England. It frequents th middle and lower shore of sandy bays, just below the surface of th sand. In colour it is sandy-greyish. It is immediately recognisabl by the extremely long arms and tiny body. The body disc measures ½ inch in diameter, the arms length being approximately 15 times the diameter of the disc, unlike other species listed where arms are about five times the diameter of the discs.

BLACK BRITTLE STAR (*Ophiocomina nigra*)
Common, on west and south coasts of England, uncommon elsewhere. It frequents the lower shore, occurring in gravelly places. Despite its name it is not always black, varies to grey, brown, sometimes pinkish-brown. The disc feels smooth, the five arms being thin and tapering, with numerous long spines. The disc is up to an inch in diameter, the arms approximately five times the diameter of the disc in length.

COMMON BRITTLE STAR (*Ophiothrix fragilis*)
Very common. It frequents the lower shore, occurring on sandy beaches, sometimes under stones or amongst various seaweeds. Variable in colour, red, purple, dark violet, yellow, pale grey, pale orange-yellow, it sometimes has red spots or bands; the five arms are grey or white banded with pink, alternatively may be completely white with one of these colours as spots and bands on both the disc and arms. An aid to identification is the body disc with five rays of spines which converge on the spiny middle. The arms are long, thin and tapering, there being seven toothed spines on the joints in the arms. It is a very frail creature, and the arms are often missing or partly broken off. The disc is about ⅘ inch in diameter, sometimes longer, the length of arms being approximately five times the diameter of the disc.

DAISY BRITTLE STAR (*Ophiopholis aculeata*) Fig. 84 (b)
Locally common, more so northern coasts, infrequent in southern England. It frequents the lower shore, where it occurs under stones, rocks, in crevices and similar. It is variable in colour, the disc is red or blue with paler markings, the five arms banded in darker hue of same colour. An aid to identification is the shape of the disc which is less round, but almost pentagonal, the points being in-between the arms. The thick, shorter arms have short spines. The disc is about ½ inch in diameter, length of the arms being up to four times the diameter of he disc.

SAND STAR (*Ophiura texturata*)
Common. It frequents the lower shore of sandy areas. The colour is variable, orange-yellow or red-brown above, sometimes spotted; being

yellowish-white or white on the underside. The five arms bear soft spines that remain flat upon the length of the arms instead of standing stiffly erect. The disc is ½ to 1 inch in diameter, the arms approximately five times the diameter of the disc.

THE SEA URCHINS

The oval, globular or heart-shaped body of the Sea Urchins (*Echinoidea*) is covered with protective, movable spines, making them easy to identify. The name was given due to their similarity to the hedgehog or hedgepig which in rural areas is called an "urchin". These sea urchins are armless, their bodies having holes or grooves out of which the tube-like feet project for several inches and give the animal a firm grip so that it is difficult to dislodge from an object. Under the spines, below a thin layer of tissue, there is a thin, brittle limy shell or "test", an inner skeleton, to protect the animal's internal organs. The shell or "test" is composed of small, tightly-fitting plates, the animal growing by increasing the size of each plate. The mouth is situated on the underside of the body and has five sharp "teeth", in a complex structure called "Aristotle's Lantern" (See Fig. 85), which is used to feed on vegetation and animal matter. Dead examples, with all the spines missing, are often cast ashore and then the beautiful patterning, coloration and construction can be seen. Only the rows of wart-like pimples indicate where the spines were formerly attached. Fossil sea urchins are frequently found in chalk or areas once covered by the sea. Some of the life cycle of the sea urchins is still unknown, but spawning takes place in Spring, the young larvae, unlike their parents in shape, living their early life freely in the plankton, before developing into the familiar spiny animal.

COMMON HEART URCHIN OR SEA POTATO (*Echinocardium cordatum*) Fig. 85

Very common. It frequents the lower shore, where it burrows in sand to a usual depth of about three inches, sometimes deeper; it is communal, several or many living together in a group. Alive it is sandy-brown but when dead the inner shell or "test" is white when it is found on the shore. The body is roughly heart-shaped, about 1½ inches broad, slightly longer, but smaller or larger examples may be found up to 3 inches in length. The soft, weak, easily-broken spines point backwards to the shape of the body. It has no teeth, the tube-feet being used to pass food obtained from the surface of the burrow into th

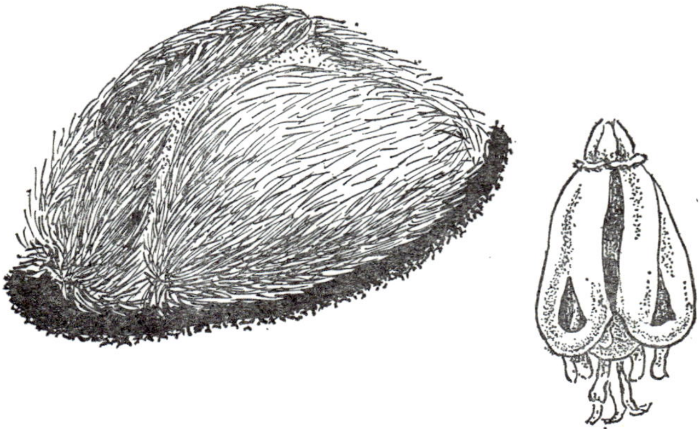

Fig. 85 Common Heart Urchin or Sea Potato
Echinocardium cordatum

mouth. It was called the Sea Potato due to a resemblance when the spines are removed.

COMMON SEA URCHIN (*Echinus esculentus*)
Common. It frequents the lower shore, especially in spring, among seaweeds. The live test, spherical in shape from above, flattened slightly, is dark red or reddish-purple, with white tubercles, the tough, blunt, furrowed spines red or white with red-purple tips. It is up to 4 inches in diameter, sometimes larger. It is edible.

GREEN SEA URCHIN OR CAKE URCHIN (*Echinocyamus pusillus*)
Common. It frequents the lower shore, on sandy and gravelly places. The live test is bright green or greyish-green, the dead test dull white. In shape it is oval, flattened, with very short spines, closely placed. The smallest British sea urchin, its body is up to $\frac{3}{4}$ inch in length.

PURPLE HEART URCHIN (*Spatangus purpureus*)
Locally common. It frequents the lower shore where it burrows in sand and gravel. Alive it is violet or purple, spines paler. The body is heart-shaped, about three inches broad, slightly longer, but larger examples may be found. The spines are weak, easily broken and point backwards to its shape of the body, those on the back being long and curved. It has no teeth, the tube-feet being used to pass food obtained from the surface of the burrow into the mouth.

PURPLE-TIPPED SEA URCHIN (*Psammechinus miliaris*)

Common. It frequents the lower shore, under stones, under and on overhanging surfaces of rocks. It sometimes covers itself with seaweed, shell fragments and small pebbles or stones, as camouflage, making it difficult to see in its habitat. Alive, from above it is almost spherical in shape, flattened, a dull or pale green, the tips of the short, tough spines being violet or purple. It is up to 2 inches in diameter.

THE SEA CUCUMBERS

The Sea Cucumbers (*Holothuroidea*) are not, despite their name and shape, marine vegetable plants, but are animals. They have long bodies, with the mouth at one end surrounded by tentacles used for feeding. The leathery or warty body has no strong internal skeleton, only plates contained in the skin. On the underside, on which the animal moves, mouth pointed forward, it has five rows of sucker tube-feet. One species burrows in sand, the others are usually found among or under stones, rocks, seaweeds. Most are tiny and difficult to discover. To avoid capture some species disgorge their digestive organs, growing new ones if they succeed in escaping.

CUCUMARIA LACTEA Fig. 86 (a)

Locally common, except on the east coast of England. It frequents the lower shore where it occurs on stony and shell-fragmented shores, also among seaweeds. The body colour is pink-brown or white; the ten tentacles are white. It has a thick, leathery, tough skin, the cylindrical body having only a few tube-feet in a line. It measures up to 1½ inches in length.

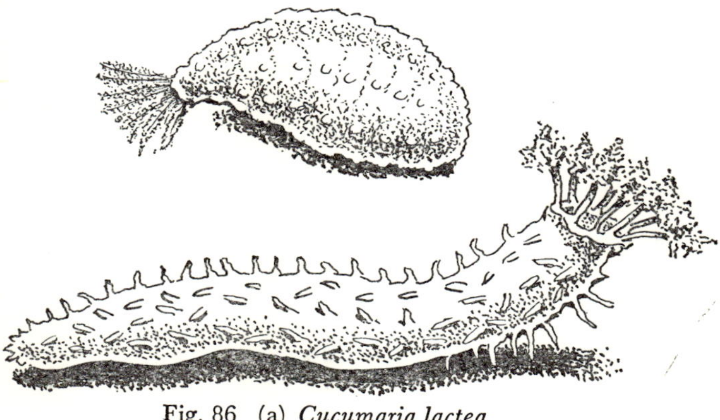

Fig. 86 (a) *Cucumaria lactea*
(b) Sea Gherkin *Cucumaria saxicola*

CUCUMARIA NORMANI

Locally common on south and south-west coasts. It frequents the lower shore, among or under stones and rocks. The body is brownish-white, but when exposed to light may be black. It has ten branching tentacles. The skin is leathery and wrinkled, the body nearly cylindrical with double rows of tube-feet. It measures up to 4 inches, but occasionally may be longer.

SEA GHERKIN (*Cucumaria saxicola*) Fig. 86 (b)

Common, chiefly on the south and west coasts of England. It frequents the lower shore among or under stones and rocks, in narrow crevices. It is white in colour, the 10 white tentacles having dark branches. It has a soft, thin, but tough skin, the body is nearly cylindrical, with five rows of tube-feet above, double rows below. It measures up to 4 inches in length, but is sometimes longer.

Chapter 12

Jellyfish

The Jellyfish (*Scyphozoa*) belong to the same group as the Sea Anemones, Sea Firs and Corals, called the *Coelenterata*. The name comes from the Greek for "hollow stomach". It is not difficult to understand why these transparent and semi-transparent marine creatures were given this name, Jellyfish, although they are not true fish. They consist chiefly of a jelly-like substance. Their body structure is primitive and simple, comprising an outer layer of cells around the body, with an inner layer lining the gut and in between the larger proportion of "jelly". The mouth, bounded by its four long trailing "lips", which may be several times divided, is the only body opening, sometimes surrounded on the disc edge by trailing tentacles, the latter armed with stinging cells. These cells have a two-fold purpose: to give protection to the fragile and otherwise defenceless body and to obtain the tiny, active sea creatures, minute plankton, fish fry, etc., on which they feed, by paralysing them and drawing them into the mouth.

Jellyfish float in the sea like a parachute or umbrella, but when seen from a pier some of the gently pulsating movements of the larger species in the water have a similarity to a graceful ballet. Others are like transparent globes passing by. These larger, radial disc or bell-shaped Jellyfish are not entirely at the mercy of the sea and when watched it will be seen that by spreading then contracting their bodies they can move with the disc pointed in a forward direction through the water. To be appreciated Jellyfish have to be seen like this in their natural element. But thousands are cast up on the shore, among rocks, on mudflats or harbour bottoms, by a rough sea or left behind as the tide recedes. Then the Jellyfish is ugly, resembling a heap or pancake of jelly. It dies quickly and having no bone structure eventually disappears entirely. The hot sun and air dries the jelly material and evaporates it. If you pass the same spot an hour or so later there may be no trace of it or, if it was lying on sand or mud, only a slight ring-shaped imprint where it had been.

The Jellyfish begins life as an egg. From these hatch tiny larvae which swim about using threads called "cilia". As they develop into what is called the "hydroid stage" the "cilia" disappear and the small, white larvae, called "polyps", with long, flowing tentacles, fix themseves to rocks in pools, weeds, or piles on piers and jetties beyond low water mark, using a type of stalk-like "foot" or disc. Now looking similar to tiny sea anemones the body elongates into a tube shape, the tentacles being withdrawn. The tube develops into a series of rings, looking rather like a stack of upturned saucers, and eventually it splits into separate pieces, each eight-lobed disc or "saucer", the "medusa", floating away to become a new Jellyfish. This type of development is known as strobilisation. After further growth the lobes of the tiny Jellyfish disappear and the familiar shape of the disc or "bell" of the Jellyfish is reached.

Stranding on the shore is not the only hazard facing the Jellyfish. During its development it is parasiticized by the larval stage of *Peachia hastata*, a burrowing anemone. The Moon Jelly or Common Jellyfish, because it is so frequently seen close to the shore, is often pelted with stones and pebbles by children, the unfortunate creature disintegrating into pieces the more it is hit. In such cases this is just a childish beach pastime, as target practice, but adults often kill these Jellyfish, too, through a mistaken belief that the Moon Jelly stings. But, unlike certain other species, this commonly-met marine animal is harmless to us and will not sting even if touched in the water while swimming. The larger types of Jellyfish also sometimes suffer the indignity of being robbed of their meals. The tiny fish fry, such as whiting, cod, horse mackerel, lurk beneath and among the tentacles and "bell" or disc of the Jellyfish, taking the food the Jellyfish would otherwise have obtained. The small fish also deliberately stay among and beneath the Jellyfish to obtain a dangerous form of protection from the attentions of larger fish and other predators.

BELL JELLYFISH OR STALKED JELLYFISH (*Haliclystus auricula*) [*Lucernaridae*] Fig. 87 (a)
Common on south-west coast of England, rare elsewhere. It frequents the lower shore and does not swim through the water but is usually attached to stones, rocks, seaweed and eel-grass. It achieves this by means of a sucker at its base, but when it requires to do so it can move to another site and then may come to the surface. In shape it is like an

upturned trumpet or bell, with the edge lobed, at the tip of each lobe there being a group of tentacles. In colour it is green or red with paler lines of the same colour. Like other Jellyfishes, out of water the Bell Jellyfish completely loses its frail and delicate beauty. It must be carefully searched for, as it measures only up to ½ inch in height.

COMPASS JELLYFISH (*Chrysaora hyoscella*) Fig. 87 (b)
Uncommon, chiefly occurring on south and south-west coasts. It has 24 long tentacles trailing from the umbrella disc's lobed margin and stinging cells able to penetrate human skin that contacts them, so do not touch it and avoid it if you meet it while swimming. In colour it is milky-white with brown areas on the disc lobes. In the top centre of the disc there is also a brown area. It was so-named because of the pattern of triangular brown and white marking which radiates out the centre of the umbrella disc, resembling a compass "rose", which is an aid to identification. It measures up to 16 inches in diameter, but between 4 to 8 inches is more usual.

CYANEA LAMARCKI
Locally common, more frequent on the east coast of England and Scotland, occasionally on the south and south-west coasts. The margin of the disc or "bell" is deeply lobed, from which hang eight tufts of long tentacles, which have poison stinging cells dangerous to human bathers who touch them. In colour it is a bluish-violet hue. Smaller than its relative Lion's Mane Jellyfish (*Cyanea capillata*) it measures up to 15 inches in diameter, but between 3 to 10 inches is more usual.

LION'S MANE JELLYFISH (*Cyanea capillata*) Fig. 87 (c)
Rare, chiefly on the south-west coast of England. Like its close relative *Cyanea lamarcki*, the margin of the disc or "bell" is deeply lobed and bears eight tufts of long tentacles which have poison stinging cells that are extremely dangerous to human bathers, if touched causing cramp. In colour it is yellow, brownish-yellow or brownish-red and measures up to 18 inches in diameter.

MOON JELLY OR COMMON JELLYFISH (*Aurelia aurita*) Fig. 87 (d)
Common. The Jellyfish most frequently seen around the British coasts, particularly in shallow, sheltered water; it also occurs in estuaries, tidal parts of rivers, salt-water inlets, creeks, in docks and harbours, sometimes in large numbers. The nearly-flat disc, like an open umbrella or

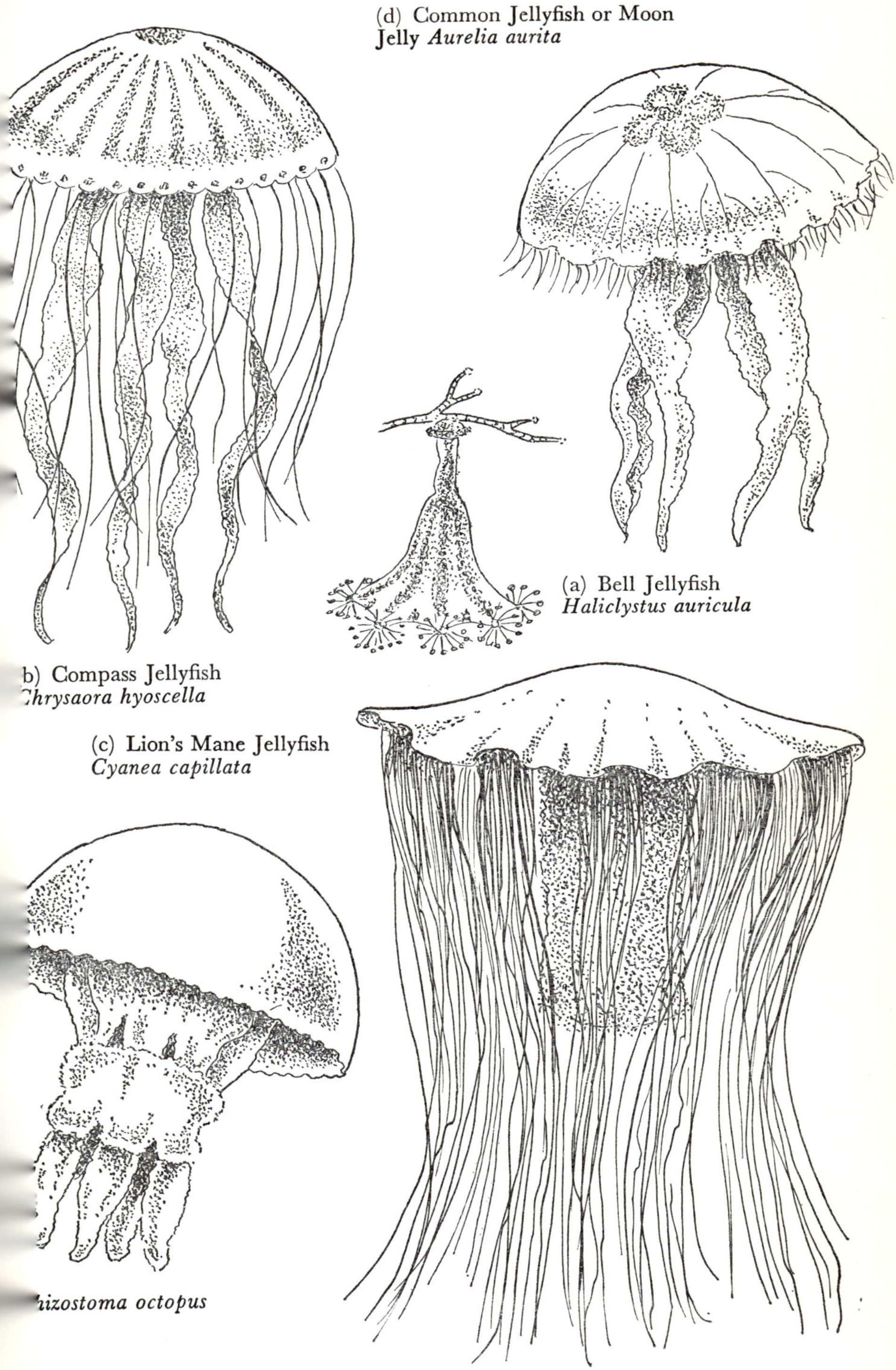

Fig. 87 Some Jellyfish

parachute, is almost colourless and transparent and has in the centre four violet or lilac "rings" or "crescent" markings. These are the site of the reproductive organs and are not its eyes, as is so frequently thought when this Jellyfish is found. Around the margin of the disc there is a frill of numerous small tentacles but its sting cells do not harm human skin and present no danger to swimmers, etc., which is fortunate, since it is so common. It measures up to 15 inches in diameter, but between 3 to 8 inches is more usual.

PORTUGUESE MAN-O'-WAR (*Physalia physalia*) Fig. 88
Locally common periodically. This is not a British Jellyfish and belongs to a different group (*Siphonophora*), but due to its similarity to coastal Jellyfish and the fact that in favourable seasons large numbers of them occur close to the British southern and western coasts, after drifting from the Atlantic, it can be included. The body comprises an elongated-

Fig. 88 Portuguese Man-o'-War *Physalia physalia*

oval shaped, gas-filled, floating bladder, about six inches in length, from which a large number of long tentacles trail. In some exceptional specimens these tentacles have measured up to 40 feet or more in length. Along the top of its body there is a raised, notched crest which acts as a "sail" and assists it in floating on the sea's surface, carrying it along to new areas, feeding on the prey which is killed by touching the sting cells on its tentacles. Curiously, however, many small fish often shelter within a Portuguese Man-O'-War, living dangerously in this way to obtain protection from other predators. It is very poisonous and should never be approached in the sea as touching a tentacle will give naked human skin an extremely painful sting. If this Jellyfish is found alive cast up on the shore remember that the poisonous tentacles can still inflict such an injury. It is light blue or pinkish in colour, the tentacles bluish, and the crest is also a pinkish hue.

RHIZOSTOMA OCTOPUS Fig. 87 (e)
Locally common, chiefly on southern and western coasts of England. This is the largest British Jellyfish and has a "solidly massive" appearance. The mouth "arms" or lobes are long and joined to form a grouped mass. It is pale green, pale blue or grey, with a darker shade of the same colour or purplish, on the lobed edge of the "bell" or disc, which has no tentacles. Because of its size it is perhaps rather frightening, but is completely harmless to human beings. It measures up to 24 inches across the disc or "bell".

COMB JELLIES (*Ctenophora*)

Superficially they resemble the true Jellyfish but the carnivorous Comb Jellis (*Ctenophores*) have several clear differences. They develop from eggs but do not have an intermediate stage, instead becoming miniature Comb Jellies direct from the egg. The transparent Comb Jellies have eight rows of fine combs or numerous, small, iridescent, vibratile swimming plates, "ctenes", which the animal, with mouth pointing forward, uses by beating to swim, unlike the true Jellyfishes. They also do not possess stinging cells. In some seasons they are very common, appearing close to the sea's surface or are stranded in pools, from spring to autumn.

BEROE CUCUMIS
Occasionally common. Larger than its relative the Sea Gooseberry (*Pleurobrachia pileus*) it is pear-shaped or thimble-shaped with a nipped-in "waist". It does not have tentacles which is an aid to identification. The rows of vibrating combs or swimming plates are clearly

visible when mobile, while the iridescent colours of the combs and its frailness give this species a delicate beauty in the water. It is an open sea animal but sometimes comes close to the shore or is found in rock pools at low tide. In colour it is transparent colourless or a pinkish hue. It measures up to 2 inches in length.

SEA GOOSEBERRY (*Pleurobrachia pileus*) Fig. 89
Occasionally common. Oval, as a gooseberry, and from this shape gets its name. It has two long, waving, branching, withdrawable tentacles, up to 9 inches in length, which it uses to lure and trap its victims. It has eight rows of vibrating combs or swimming plates and is beautiful when progressing through the water. It can be found near the shore or when stranded by the receding tide in pools. It measures about ½ inch in width and up to ¾ inch in length.

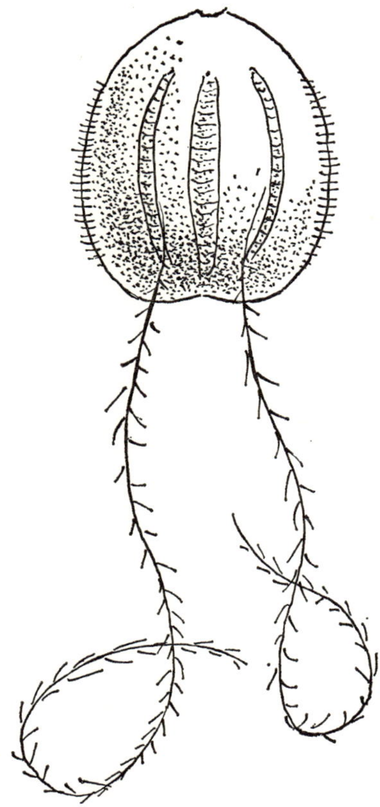

Fig. 89 Sea Gooseberry *Pleurobrachia pileus*

Chapter 13

Sponges, Sea Mats and Sea Squirts

SPONGES

The Sponges (*Porifera*) appear to be plants and were classed as such by the early naturalists. They are, however, many-celled animals. Their body structure is simple, being strengthened with a skeleton of tiny, limy splinters or spicules of a siliceous material. This can be felt when some examples are squeezed and the water exuded. As soon as released this type usually absorbs water again. The walls of the body have many small openings or pores through which a regular current of water is drawn, by means of whip-like "threads", into a system of small channels. From this water microscopic food particles and oxygen are filtered, the waste and water then travelling along larger channels to be discharged through one or more larger outlet surface apertures, not a mouth or mouths, called the "osculum."

Sponges reproduce in two ways (1) by eggs, which form in the wall and later hatch into minute larvae. These, after emerging, seek a habitable site for themselves and develop into Sponges; (2) by the parent Sponge forming "buds" which are liberated.

None of the 200 or so species of Sponges living on the lower shore around the British coasts and estuaries are as large as, or similar to, the familiar bathroom Sponge which thrives in warmer seas than ours. Most British species are small, although some are Sponge-like in texture. All are attached to another object.

There are three main types, – flattened; tubular; or crust-like. The Purse Sponges, fixed at one end to an object, have shaped projections and are known as "vases" due to their apparent similarity to these, although sometimes they look more like straight fingers on a hand. The second group are the Boring Sponges, which obviously penetrate into shells and limestone. The third group are the Encrusting Sponges, which form a large or small area of sponge-like or "crust" material upon the habitat. The colours and shape of the Sponges, even in members of the same species living close together on a habitat, can vary considerably and identification of the Sponges is thus not an easy task. Only some of the common species likely to be met with are described.

BLOOD-RED SPONGE (*Hymeniacidon sanguinea*) Fig. 90 (a)
Common. It frequents the middle and lower shore. It is an Encrusting Sponge occurring on or under seaweeds, in crevices and surface undersides of jutting rocks, also in pools on rocks and stones and in muddy estuaries. Its shape varies considerably, circular or irregular, sometimes in small fragments or may be a mass covering an extensive area of several feet. The surface is roughened by grooves and its pores. Colour, despite the name, is also variable, usually blood-red, orange, or scarlet.

BREADCRUMB SPONGE (*Halichondria panicea*) Fig. 90 (b)
Common. It frequents the lower shore. It is an Encrusting Sponge occurring on or under seaweeds, in crevices and sheltered surfaces and undersides of jutting rocks and other objects. Its shape varies considerably, circular or irregular, sometimes in small fragments or maybe a mass covering an extensive area of several feet. It usually forms lumps on seaweeds and flat fleshy encrustations on stones and rocks. Its surface is almost smooth, with a texture and appearance that gives it its name, except that it is patterned with raised openings of its pores resembling the craters of tiny volcanoes. Its colour is also variable, being dirty white, grey, green, yellow, orange, brown, with several different colours of the same species living closely together. After it has been dead for a time the Sponge becomes white or greyish.

CLIONA CELATA Fig. 91
Common. It frequents the lower shore. It is a Boring Sponge found o shells. If the Sponge is absent its previous presence may be seen b many small, round holes excavated into bivalve and univalve shells This Sponge is a pest in oyster beds. In some examples holes are ver numerous. It also bores into limestone rocks. A tiny lump of the livin Sponge may be seen inside each hole. Its colour is a pale yellow or pal orange.

GRANTIA COMPRESSA Fig. 90 (c)
Common. It frequents the lower shore where it occurs on the slopin sides of boulders and under jutting parts of rocks and stones, als frequently on or among some of the red seaweeds. An aid to ident

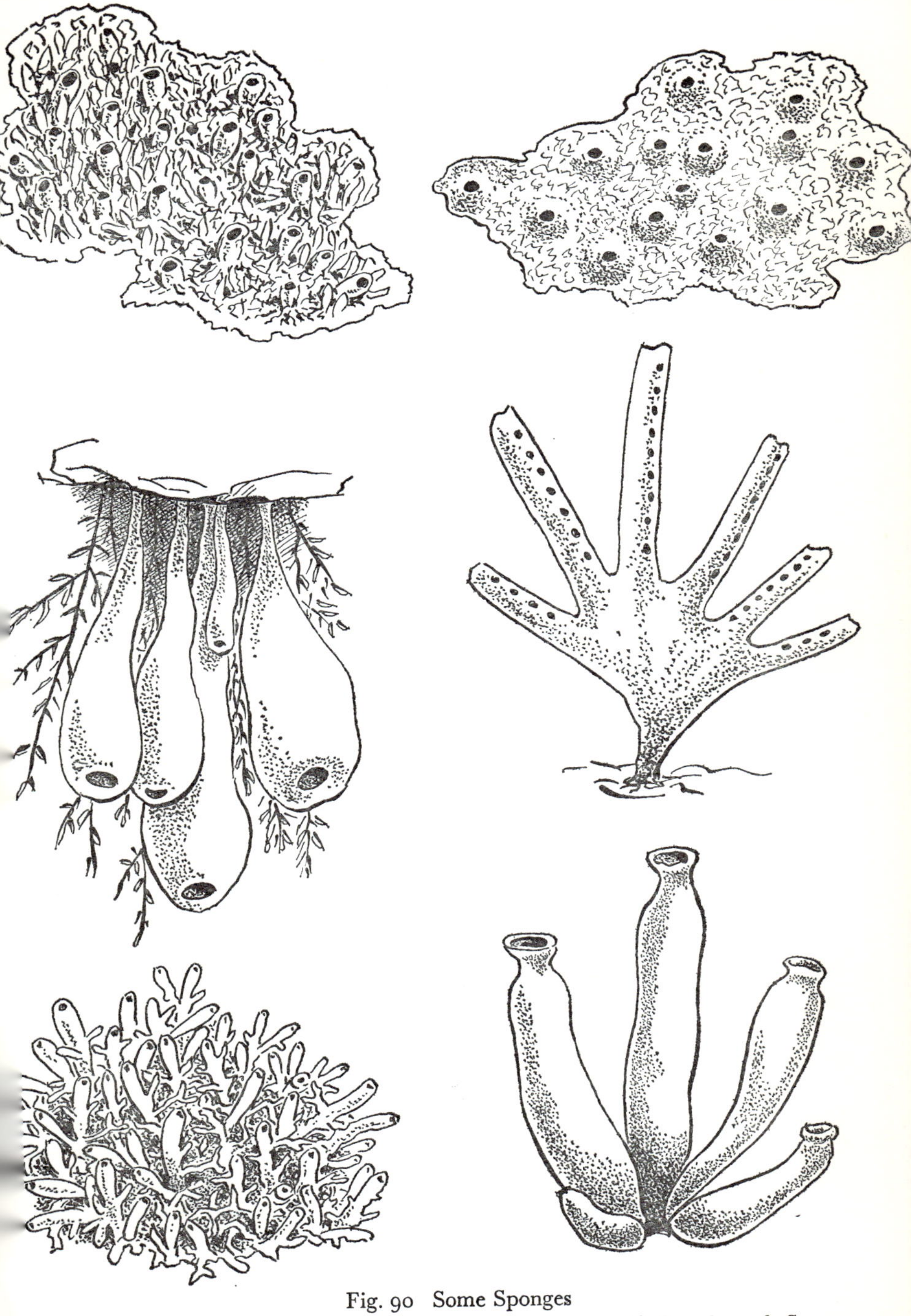

Fig. 90 Some Sponges

(a) Blood-Red Sponge *Hymeniacidon sanguinea*

(b) Breadcrumb Sponge *Halichondria panicea*

(c) *Grantia compressa*

(d) *Leucosolenia botryoides*

(e) Mermaid's Gloves Sponge *Haliclona oculata*

(f) *Sycon coronatum*

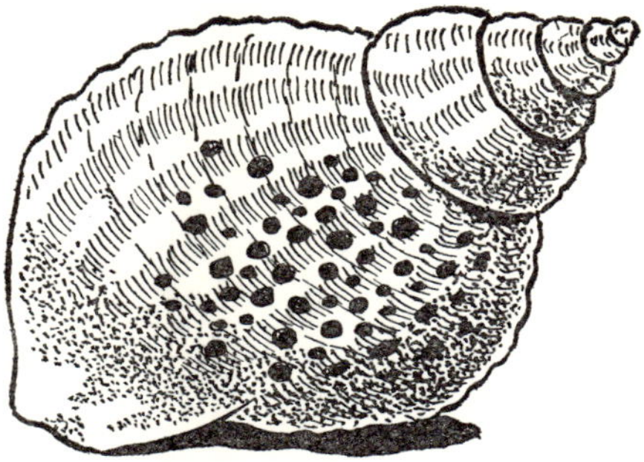

Fig. 91 *Cliona celata*

fication is that it hangs downwards. It is a flattened Purse Sponge, long and narrow, but more usually oval in shape, attached to the object by a short stalk. It measures up to two inches in length. Several examples may be found together. The colour is pale brownish, yellowish, or greyish.

LEUCOSOLENIA BOTRYOIDES Fig. 90 (d)
Common. It frequents the middle shore and upper lower shore, where it occurs in pools on seaweeds. It is a Purse Sponge with stiff, projecting branches, up to $\frac{1}{2}$ inch in height, which sometimes have a similarity to upright bunches of tiny bananas. The colour is usually white, though maybe other pale shades.

MERMAID'S GLOVES SPONGE (*Haliclona oculata*) Fig. 90 (e)
Common. It frequents the lower shore where it occurs on seaweeds and rocks. It is a Purse Sponge, with upright, projecting, numerous stiff branches, the sides of which have many small openings. It measures up to 12 inches in height. It gets its name from a supposed resemblance of the branches and their texture to the fingers of a hand wearing a glove. The colour is whitish or greyish.

SYCON CORONATUM Fig. 90 (f)
Common. It frequents the middle shore, where it occurs in sheltered rock pools, on and under stones and debris and among the seaweeds. It is a Purse Sponge, with several stiff, upright, cylindrical shaped projections or "vases". Sometimes these are separate or developing from one short stalk. The hole or osculum at the top of each "vase" has a fringe of long, bristle-like spicules. The colour may be a sandy-yellow or greyish-yellow, but if it is rubbed against another object by the waves

this is removed to reveal a light grey colour underneath. It is up to 2 inches in height.

SEA MATS

The Sea Mats are members of a group of animals called the *Polyzoa* ("many animals") or *Bryozoa* ("moss animals"). They are formed of colonies of numerous individual animals, which gave rise to their name *Polyzoa*. Like the Sponges, the form the colonies take is similar to the flat fronds of seaweeds or a branched plant shape or jelly-like masses, although most of those discovered on the lower shore are irregularly-shaped encrustations upon rocks, stones, shells, or large seaweeds, hence the name "Sea Mat". The encrusting layer or plant-like delicately-beautiful colony comprises numerous, small, separate, cell-like, limy or siliceous compartments, each of which contains a minute animal. A pocket lens is usually needed to see this construction. Most of these animals have a U-shaped fringe of withdrawable tentacles around the mouth, these tentacles bearing numerous cilia or hairs, which vibrate to create a flow of water into the mouth. In this way the animal obtains tiny food particles and oxygen. The extension of the tentacles by each animal while obtaining food gives the entire Sea Mat a whitish, mass-like appearance, from which originated their name. The animals reproduce themselves either by budding or producing eggs which hatch into minute larvae. The latter join the plankton, but eventually descend to a suitable habitat to begin a new colony.

ALCYONIDIUM HIRSUTUM Fig. 92 (a)

Common. It frequents the lower shore where it occurs on seaweeds, especially the stiff red seaweeds. It is regularly washed ashore with seaweed. The colony shape is a mass of sponge-like, gelatinous material with fleshy, irregularly lobed and blunt projections like fingers. The colony size measures up to 12 inches in length. Its colour varies from greenish, greyish or a reddish-biscuit hue.

BUGULA TURBINATA

Common. It frequents the middle shore and lower shore, where it occurs on rocks, mainly on the overhanging rock ledges, where it is draped in large numbers. It also occurs among red seaweeds and purse sponges. The colony shape is numerously branched and tufted. Its size may measure up to 2 inches in length. The colour is a light orange-yellow with darker orange areas.

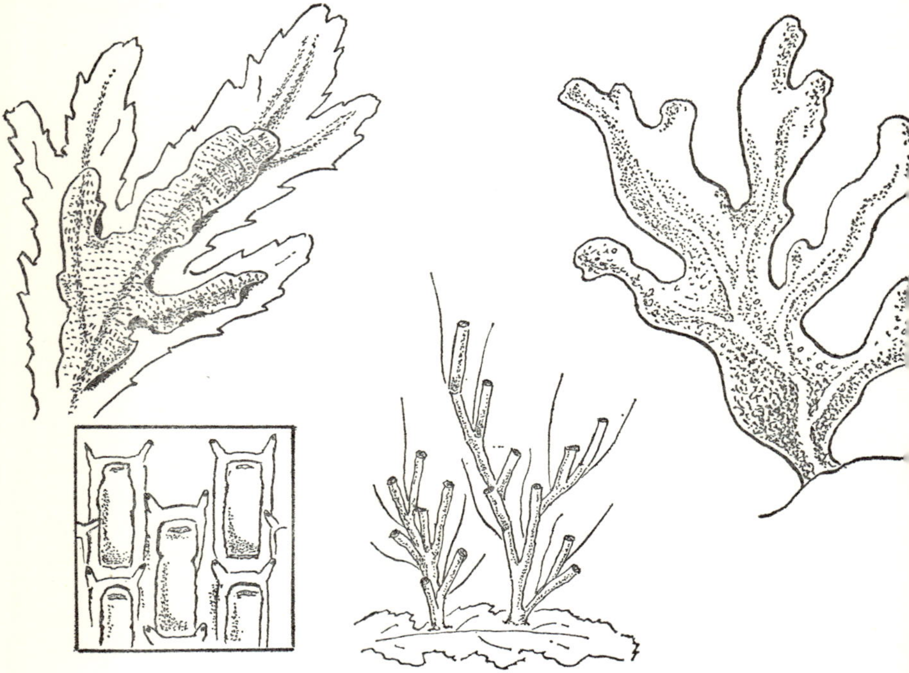

Fig. 92 Some Sea Mats

(c) *Membranipora membranacea*

(a) *Alcyonidium hirsutum*

(b) *Crisia eburnea*

CRISIA EBURNEA Fig. 92 (b)
Common. It frequents the middle shore and lower shore where it occur on rocks, stones, seaweed, also shells and similar. The colony shape i numerously branched tufts, the branches comprising curved tubes, wit a long spine behind each tube. The colony size is up to an inch in heigh Its colour is whitish.

FLUSTRELLA HISPIDA
Common. It frequents the middle shore and lower shore where it occu on the lower parts of the fronds and stems of seaweeds, especiall carrageen moss and the wracks. The colony shape is an irregular, e crusting mass. An aid to identification is the numerous red-brow horny, protective spines which give it a coarse "hairy" appearanc Colony size varies, up to 18 square inches. The colour is dull greyis brown or dark red-brown.

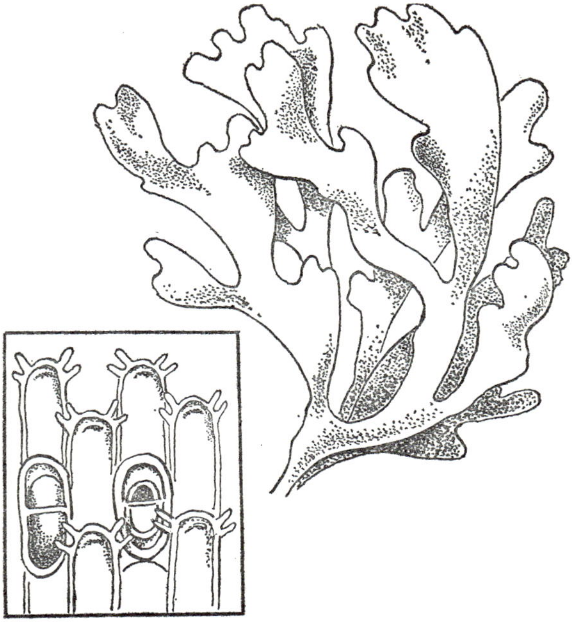

Fig. 93 Horn Wrack *Flustra foliacea*

HORN WRACK (*Flustra foliacea*) Fig. 93

Very common. It frequents deep water below the lower shore, where it occurs thickly covering a wide area, but portions or complete examples are often found washed ashore, particularly after a storm. The colony shape is numerously divided, lobed, flat, branched fronds, with rounded, broad ends. Because of this it could easily be mistaken for a seaweed but examination with a lens will reveal the arrangement of tiny, rectangular compartments, each one having an animal occupant. The colony measures up to 6 inches in height. The colour, when living, is brownish, but after death and being cast ashore, where it becomes dry and brittle, it is a greyish-yellow or straw colour.

MEMBRANIPORA MEMBRANACEA Fig. 92 (c)

Very common. It frequents the middle shore and lower shore, also in brackish water, where it occurs on seaweeds, particularly oarweeds and wracks, also on shells, stones and other objects. The colony shape is an irregular, encrusting mass. Its size also varies and may be up to 18 square inches, covering a large area of the seaweed with its lace-like growth. The colour is white.

UMBONULA VERRUCOSA

Fairly common. It frequents the lower shore where it occurs on seaweeds, particularly oarweeds, rocks, stones and other objects. The colony shape is also an irregular, encrusting mass, but it is thin and fragile, its size being much smaller and usually only 3 inches or less across. Its colour is a pinkish or rosy red.

SEA SQUIRTS

The Sea Squirts (*Tunicata* or *Ascidiacea*) are members of the group *Chordata,* which includes the backboned vertebrates – birds, animals, fish, reptiles, even man. The reason why the Sea Squirts or *Tunicata,* which have no backbone, are included is that their life cycle includes a "tadpole-like" stage with the vestige of a backbone, the notochord, and other embryonic features that have a similarity to the first stages of their distant relatives, the vertebrates.

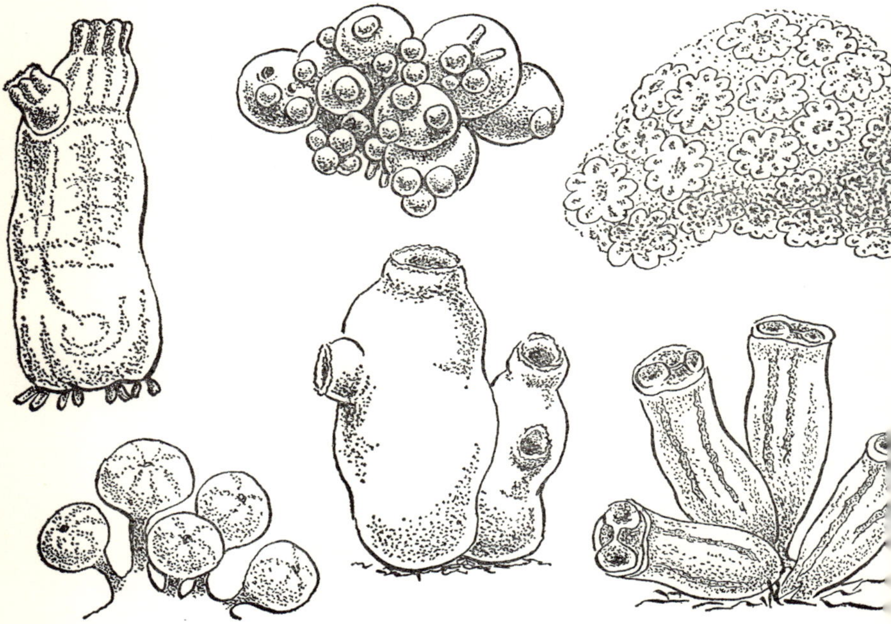

Fig. 94 Some Sea Squirts

(f) Tube Sea Squirt *Ciona intestinalis*

(c) Gooseberry Sea Squirt *Dendrodoa grossularia*

(e) Star Sea Squi[rt] *Botryllus schlosseri*

(d) *Morchellium argus*

(a) *Ascidiella aspersa*

(b) *Clavelina lepadiformi[s]*

The long, oval, rounded, sometimes flat, jelly-like or transparent, soft, sac-like body of the mature Sea Squirt is contained inside a protecting, tougher, leathery mantle or "tunic". It has two openings or siphons. One of these is the mouth drawing into the body a flow of water, which is strained to obtain a supply of fine food particles and oxygen, the water and waste material then being expelled via the second or anal siphon. When squeezed by the fingers or a thumb the water squirts from the siphons.

There are three types of shore Sea Squirt. Some of them are solitary, separate living animals; others live solitary but are connected to other individuals by a network of creeping roots or are joined at their base, while a third type live together in a jelly-like area as an encrusting colony. For this reason and because they feed and exist in a similar manner they are included in the same chapter as Sponges, for which they may be mistaken, and Sea Mats, although in other respects they are very different. Sea Squirts are found attached by all or part of one side or the base of their body to rocks, stones, seaweeds, though a few examples live in mud.

The Sea Squirts produce eggs which hatch into minute "tadpole-shaped" larvae. These swim among the plankton until they descend to a suitable habitat where they attach themselves and mature into adults, during which development the notochord disappears.

ASCIDIA CONCHILEGA

Common. It frequents the lower shore where it occurs on shells, stones, rocks and other objects. It lives as a separate animal. The body is oval-shaped with a rough surface and adheres to an object by one side. It has a short siphon at one end, but the other larger siphon is situated about two-thirds of the body length apart. The skin is semi-transparent, and the blood vessels, which are green, can be seen. It measures up to inches in length.

ASCIDIELLA ASPERSA Fig. 94 (a)

Locally common. It frequents the lower shore on south and west coasts of England, also in estuaries, where it occurs on stones, rocks, harbour installations, piers, jetties and similar objects. It lives as a separate animal. The body is oval-shaped with a rough surface and adheres to an object by its base. It has a short siphon at one end, the other siphon being situated about one-third to half of the body length apart. The colour is usually a brownish or dark grey. It measures up to 3 inches in height.

BOTRYLLOIDES LEACHI

Common. It frequents the lower shore and occurs on rocks, stones, seaweeds. It lives as a colony on the habitat, forming a thin, flat, jelly-like encrustation. Around its siphons there are individual animals forming long patterns of waved lines, occasionally broken into sections. The colours of the colony vary, yellow, orange, brownish or greyish, sometimes several variations being in close proximity. The colony area measures up to 5 inches across.

CLAVELINA LEPADIFORMIS Fig. 94 (b)

Common. It frequents the lower shore, where it occurs on stones, rocks, harbour installations, piers, jetties and similar objects. It lives as an individual animal, except that several are situated close together as a colony and share the same creeping base to adhere to the habitat. The soft, transparent body is long, narrow, stalked. The two siphons are sited at the upper end of the body, the largest siphon is at the top, the smaller siphon slightly below it. The body comprises several colours usually yellow, white, pinkish, orange, brownish, and measures up to 2 inches in height, although an inch is more usual.

GOOSEBERRY SEA SQUIRT (*Dendrodoa grossularia*) Fig. 94 (c)

Very common. It frequents the lower shore where it occurs on rocks under the overhanging clefts, on sides of stones, on shells, in oyster beds also attached to the stalks of seaweeds. It lives as a separate, solitar individual but also occurs in large numbers as a colony when they shar the same base. The body is rounded in shape with a wrinkled, leather appearance. The two siphons are sited at the upper end of the Se Squirt, one at the top and the other siphon slightly below it. They ar usually red in colour. The body colour is a deep purplish-red on th upper-parts and brownish-red or brownish-yellow on the lower portio Due to its colour and shape it has a supposed resemblance to a re gooseberry, hence its name. It measures up to ½ inch in height.

MORCHELLIUM ARGUS Fig. 94 (d)

Locally common on south and west coasts of England. It frequents th lower shore, also in estuaries, where it occurs under overhanging rock and sides of stones. It lives as a flat or lumpy colony closely joine together in tough, thick, stalked bunches. The individual animal is re and yellow with reddish spots. The colony is up to 1½ inches in length.

STAR SEA SQUIRT OR GOLDEN STARS (*Botryllus schlosseri*) Fig. 94 (e)
Very common. It frequents the lower shore where it occurs on rocks, stones, and seaweeds. It lives as a colony on the habitat, forming a thin, flat, jelly-like encrustation. The individual animals, each with its own inhalation siphon, form star-like patterns, every "star" having up to 12 "petals", individuals sited around a central siphon used communally by all the animals in each "star" for exhalation. Because of the unmistakable pattern it was given its name. The colours of the colony are variable, the jelly-like encrustation being pale or dull yellow, orange, light brown, greenish-blue, violet, with the individual animal "stars" vivid yellow or reddish. See BOTRYLLOIDES LEACHI. The colony area measures up to 6 inches across.

TUBE SEA SQUIRT (*Ciona intestinalis*) Fig. 94 (f)
Very common. It frequents the lower shore where it occurs on or under stones, in clefts in rocks, on harbour installations, pier piles, jetties, also undersides of buoys and ships' hulls. It lives as a solitary, separate animal. The soft, semi-transparent body is long and cylindrical. It adheres to an object by its base or lower side. The siphons are sited at one end of the body; one on the tip, the other slightly below it. The colour is variable, either yellow, light green, grey, with some red spots close to the siphons. It measures up to 6 inches in height.

Chapter 14

Sea Anemones, Corals and Sea Firs

SEA ANEMONES

The Sea Anemones (*Anthozoa*), with the Corals, Sea Firs (*Hydro-zoa*), and Jellyfish, belong to the group *Coelenterata*. They do not hav a skeleton, their body structure being simply a jelly-like sac or tube called the column, with one opening, the mouth, leading to the stomach into which small plankton creatures are passed after being paralysed or contact with the poisonous barbed stinging cells of the tentacle surrounding the mouth entrance. After digestion the victim's remain are expelled from the mouth.

Around the mouth, at the top of the column, is the area called th "disc". The base or "foot" of the sea anemone comprises a sucker-lik material which allows it to grip tightly an object on which it is living Attempts should not be made to dislodge an anemone from rock, etc., as it will be injured or destroyed. They are not permaner on the same site, however, but move slowly to another suitab habitat.

Sea anemones have two forms of life cycle. One method is fro eggs that grow into small polyps, which are emitted into the water retained inside the parent anemone until they are mature for releas when they will be able to swim about until they are finally ready anchor themselves and grow. The second type of reproduction is division of the parent's body, separating into two or more parts, or the splitting off of portions of the base or "foot" which develop in small polyps.

Although they have a resemblance to flowers – not anemones some examples are beautifully patterned and coloured when fully e panded – they are in fact carnivorous animals. Out of wat exposed by the receding tide, they become formless blobs of colour jelly.

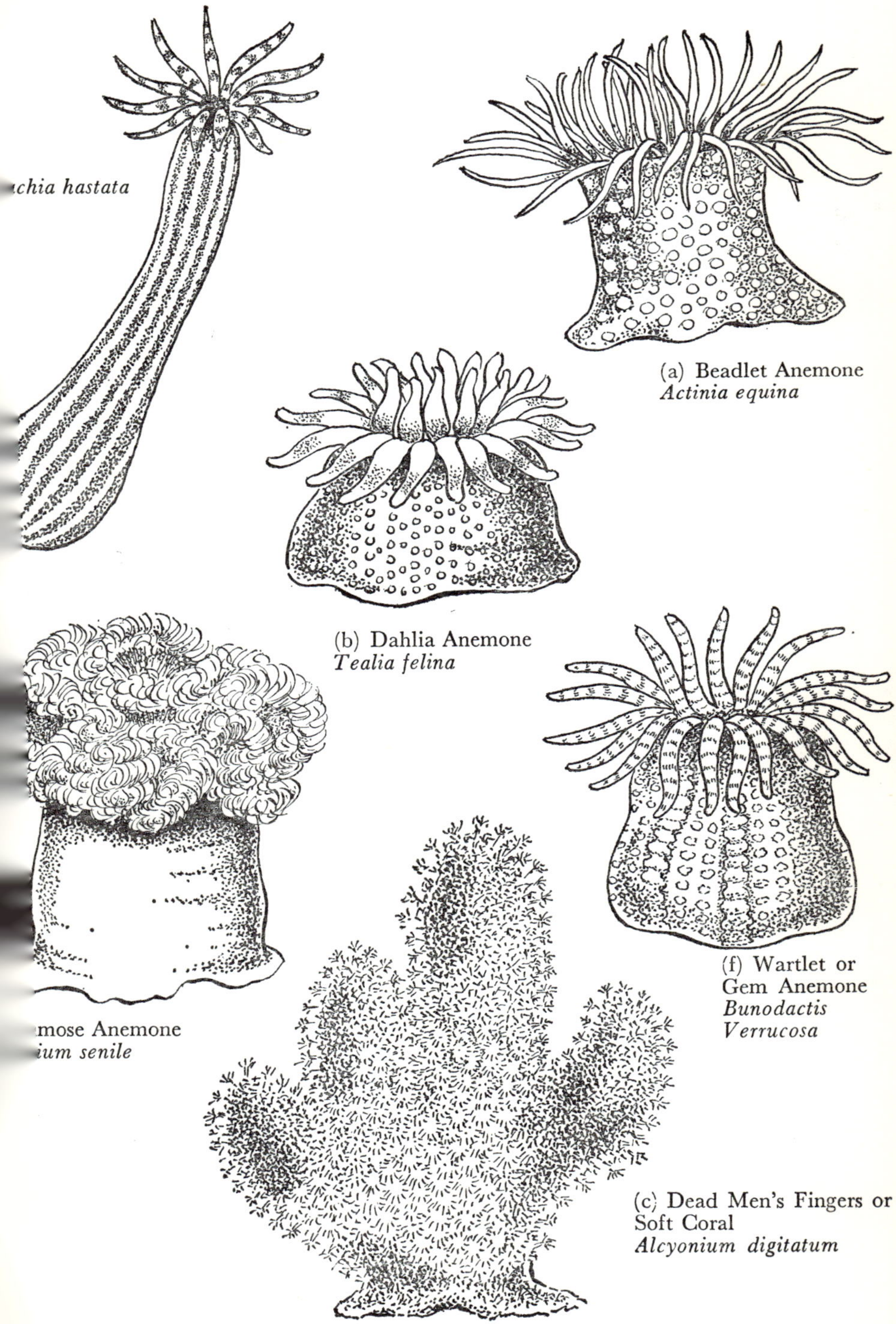

Fig. 95 Some Sea Anemones

ADAMSIA PALLIATA

Common. It frequents the lower shore where it associates with hermit crabs, adhering to the shell in which the crab is living, the base of the anemone being wrapped around the shell. This anemone has found an ingenious way of overcoming the necessity of finding a larger shell and crab companion when the anemone grows too large for the shell, by secreting horny material under its base to extend the shell size itself. This in turn ensures that the shell is continually big enough to accommodate the hermit crab as it, too, grows and would otherwise have to obtain a larger shell home. In this way the anemone and crab remain in partnership for most of their lives. The column of the anemone is pale buff and pinkish-purple, becoming cream coloured near the white tentacles and disc. It is also ornamented with gaudy magenta markings.

BEADLET ANEMONE (*Actinia equina*) Fig. 95 (a)

Very common. It frequents the middle and lower shore, in tidal pools and on rocks. It is rounded and fat, the column being up to 1½ inches in height. Widely varied in colour, usually reddish-brown, but sometimes also brown, red, yellow, green, brownish-green, yellowish-brown or strawberry red spotted with green. It has 24 blue spots around the inside of the top of the column and up to 200 tentacles are arranged in six circles. When touched the tentacles are immediately withdrawn.

DAHLIA ANEMONE (*Tealia felina*) Fig. 95 (b)

Very common. It frequents the middle and lower shore, preferring parts of pools and hiding among rocks, stones and seaweeds in the shade Fat, squat, rounded it may be several inches across at the base the column rising only an inch or two in height. The column sometime has numerous colours, variations of olive-green, pale grey or reddish occasionally with red blotches. The column also has numbers of stick grey warts to which shell, gravel and weed fragments become attached disguising the anemone. The tentacles are pale pinkish, greyish or othe shades.

DAISY ANEMONE (*Cereus pedunculatus*)

Common. It frequents the lower shore, in rock pools, crevices in rock and on sandy or muddy coasts, from which it only slightly projec above the level of the surrounding surface, but if touched it immediatel closes and withdraws into the mud or rock crack. The column is u to 4 inches in height, slender or sometimes widening into an irregul trumpet shape. The colour varies in shade, greys, browns, pinkish-re violets, cream, with lighter spots. The several hundred tentacles are either plain or spotted.

DEAD MEN'S FINGERS OR SOFT CORAL (*Alcyonium digitatum*) Fig. 95 (c)

Locally common. It frequents the lower shore, in rock pools, in the shelter of stones, pier piles, stone jetties, etc, and occurs in colonies forming circular areas. Unlike sea anemones, which have no skeleton, these colonial animals, halfway between anemones and corals, have their rounded or lobed masses of jelly-like substance strengthened by imy spicules inside it. The height of the colonies is up to 6 inches. Out of water they appear flesh-coloured and supposedly have the shape and exture of dead men's fingers, the origin of their name. In water the iny colourless polyps, each with eight feathery tentacles, emerge and add a delicate beauty. The colour of the colony is usually pure white, ellow, pinkish or orange.

EWEL ANEMONE OR GLOBEHORN (*Corynactis viridis*)

ocally common on west and south coasts. It frequents the lower shore, mong crevices, rock overhangs and similar surroundings in pools and ocky shores. Examples grow closely together in these habitats, so their olumns sometimes seem joined, but they are separate animals although ey appear to be a colony. The small, thin and frail columns are only pproximately ½ inch in height and ¼ inch across. The colour is usually rass-green, but sometimes pink, scarlet, orange, yellow, grey or white. he pale or transparent tentacles have swollen dark knobs on their tips. differs in structure from the other Anemones and is more closely lated to the corals.

PELET ANEMONE OR SNAKELOCKS ANEMONE (*Anemonia lcata*)

ommon, but restricted to west coasts and parts of the south coast. It quents the middle and lower shore, where it prefers shallow, placid, tle disturbed, sunny pools, frequenting cracks and crevices on the ttom in the brightest light, also on rocks. The column is up to 1½ hes in height, sometimes taller. The colour is usually entirely dull own or dull green, sometimes apple-green, purple-brown or pink. It s about 200 two-inch long sticky tentacles the same colour but with uve tips. These cannot be fully withdrawn by the anemone even en touched, but continue to wave about snake-like, hence its name.

PEACHIA HASTATA Fig. 95 (d)

Uncommon. It frequents the lower shore where it does not attach itself by a basal disc to objects, but creates a burrow in sand and mud up to a foot in depth where it remains for protection, projecting above the burrow only when feeding. The column is up to 5 inches in height, with a round, bulbous base. It is brownish-pink with twelve vertical lines, sometimes with brown areas around the base. About a dozen, long, pale tentacles are marked with yellowish-brown. It produces eggs that hatch into polyp larvae which, for a part of their life cycle, live as parasites upon jellyfish.

PLUMOSE ANEMONE (*Metridium senile*) Fig. 95 (e)

Common. It frequents the lower shore, in shallow rock pools, also o pier piles, stone jetties and other harbour installations, under stone and overhanging rocks or similar sheltered situations. The column i up to 5 inches in height, squat, rounded and slimy. It is variable i colour, white, cream, olive, pale brown, orange, or pink. The tentacle are very numerous, slender, almost transparent giving the anemone soft plumed or feathery appearance around the wide disc, hence it name. They are also variously coloured, sometimes matching the column but in other examples different.

SAGARTIA ELEGANS

Common. It frequents the middle and lower shore, in pools, also o rocky shores in crevices and attached to vertical and the underside c projecting rocks. The column is varied in shape, from fat, squat an round to upright and trumpet-shaped; up to 3 inches in height. The are very varied colours and patterns, so they resemble bright flower There are several varieties: *var. Miniata* is brown or cream with co trasting coloured rings or radial patterns; *var. Rosea* has a dull oran disc and pink or red-purple tentacles; *var. Nivea* has a white disc an tentacles, the column being another colour; *var. Venusta* has orange-r or yellow-brown column and disc, with whitish tentacles. The colun also emits white stinging threads if touched or interfered with.

WARTLET OR GEM ANEMONE (*Bunodactis verrucosa*) Fig. 95 (f)

Locally common on west, south-west and south coasts, in shallow poo under stones, sometimes partly hidden in a crack in the pool botto The column is up to two inches in height. It is pink or reddish w rows of large conspicuous white wartlets or "gems", while the disc and about 50 tentacles are greenish.

CORALS

Although Corals (*Anthozoa*) are normally thought of as occurring around tropical islands, where they form beautiful, massive reefs, they do exist elsewhere but do not create the dangerous reefs. Two examples are found on the coast of south-west Britain, but are not common. They differ from anemones by possessing a limy skeleton in the form of an underlying cup, the latter perhaps only evident when the coral contracts. Like the anemones and sea firs the corals feed upon minute marine creatures.

DEVONSHIRE CUP CORAL (*Caryophyllia smithii*) Fig. 96

Locally common on the south-west coast. It frequents the lower shore, on rocky places, in pools and on rock faces exposed at low water. The limy skeleton cup is $\frac{3}{4}$ inch across, $\frac{1}{2}$ inch in height, usually white or pink. The animal, coloured pinkish or chestnut, sometimes inflates itself so that it overflows the cup and resembles an anemone. The tentacles have brown markings and whitish tips.

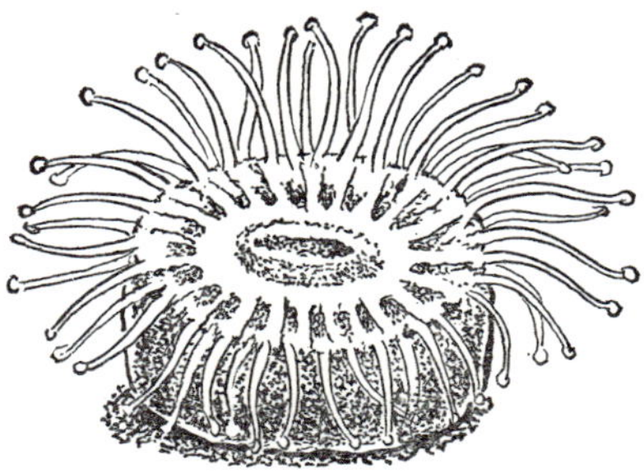

Fig. 96 Devonshire Cup Coral *Caryophyllia smithii*

STAR CORAL (*Balanophyllia regia*)

Very rare, on south-west coasts. It frequents the lower shore, usually in deep rock pools. Approximately the same size as the Devonshire Cup Coral. The animal is bright red or brilliant orange, the tentacles being transparent but marked with yellow areas, the tips being transparent.

SEA FIRS

The Sea Firs or Hydroids (*Hydrozoa*), although resembling certain types of sea plant, are in fact animals which feed on minute sea creatures they capture. They are formed into colonies, each tiny animal connected to the others by limy tissue bearing numbers of polyps and having a mouth surrounded by tentacles. The colony is developed either by sexual reproduction or asexual budding. The adult animals, medusae, in some species, break off and swim away to distribute the fertilised eggs which eventually become new hydroid colonies living on the bottom of the shore or pool, fixed on seaweeds, stones, rocks and shells.

There are two types of hydroids – one has "cups" into which the polyps withdraw for protection; the second type are "naked" or cupless. Some species have horny stalks. Most delicate are those feather or fern shaped. It is these species which have been gathered commercially, washed, dried, dyed or stained greens, purples, reds, etc., and sold in bunches for floral decoration purposes under the better-known name of "white weed" or "sea fern". There are a large number of Sea Fir species, the following being some of those most frequently to be found. Because they are rather inconspicuous a pocket lens is essential to examine them closely under water while the animals are expanded.

CLAVA SQUAMATA Fig. 97
Common. It frequents the middle and upper lower shore, in pools and

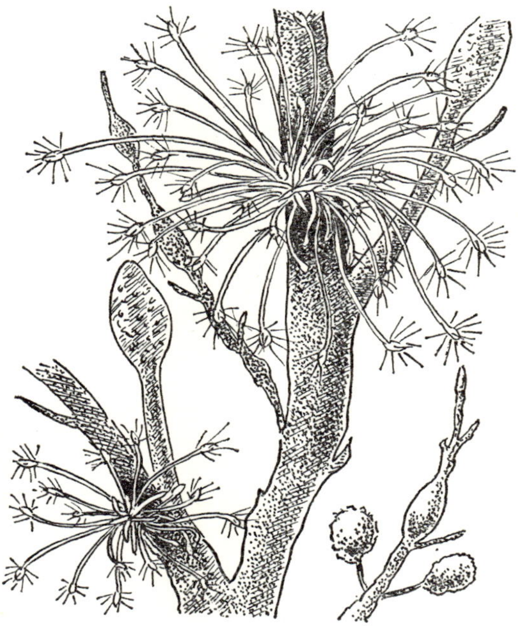

Fig. 97 A Sea Fir *Clava squamata*

on brown seaweeds, such as knotted wrack and other wracks. It occurs in pinkish or reddish crowded clusters and is cupless, having threadlike tentacles. It does not produce swimming medusae. Measures up to an inch in height.

HYDRACTINEA ECHINATA

Common. It frequents the lower shore or in same habitats where its host Crab is found. It forms a pink or brownish area on a snail shell, usually one inhabited by a hermit crab, being a partnership of mutual benefit. The Hydroid polyp extends over the shell entrance making it larger for the crab to grow into, so that it does not have to remove to a bigger shell for protection. The crab aids the Hydroid in finding food which is disturbed as the crab crawls about. This Hydroid has thread-like tentacles and is cupless, measuring only ¼ inch in height.

SEA OAK HYDROID (*Dynamena pumila*)

Very common. It frequents the middle and lower shore, on rocks, stones, shells and seaweeds, such as wracks. It sometimes occurs only as straight, stiff stems, up to ½ inch in height, or as stems bearing numerous opposite branches, with branchlets, up to 2 inches in height. The small cups are

Fig. 98 *Tubularia larynx*

in opposite pairs on every branch and stem. It is pale, almost colourless, or dirty grey. It does not produce the swimming medusae stage. Relatives of this Hydroid are gathered commercially as "sea fern".

TUBULARIA LARYNX Fig. 98
Common. It frequents the lower shore, on rocks, stones, and large brown seaweeds. It has branched stems which are buff, pale brown or yellow, the cupless "flower-heads" or zooids being light red or brownish-red, with an outer ring of drooping white tentacle threads, the inner ring of white tentacle threads being much shorter. It measures up to 2 inches in length.

Chapter 15

Seaweeds

A weed has been described as a wild plant that is growing in the wrong place, where it is unwanted, such as a garden. The sea "weeds" certainly do not fit this description, as they occur only in places for which they are ideally suited and have a vital purpose, such as preventing or slowing erosion of the shore and rocks by the sea and weather.

The group name for these marine plants is the *Algae*. You will never find a seaweed bearing flowers because they reproduce themselves asexually or sexually by means of minute spores, which are emitted and mingle with the plankton, providing a food supply for other small organisms which are themselves the prey of larger marine creatures. Those that survive, like the spores produced by fungi, must find a suitable habitat to which they attach themselves and develop from sporelings into mature plants, the fronds of the species varying widely in shape and size. It will be noticed that seaweeds do not have penetrating roots, which would be unable to obtain food from some of the habitats anyway and would also be useless to provide secure adhesion to the rock, stone, shell, other seaweed, etc. They are instead attached by a sometimes branched, anchor-like disc called the "holdfast", an apt name, which serves this task only, as all food required is secured from the sea water.

Generally, the seaweeds occur in three different zones – the green seaweeds on the upper shore and in pools, the brown on the middle to lower shore, and the red on the lower shore, but there are variations of this depending on local conditions. The visitor to the shore will soon notice that the greens tend to be frail and simple seaweeds, the browns are much tougher, coarser, rubbery or leathery in texture, ideally adapted to the pounding of the waves, while the reds again are delicate, although some of the small examples are more like coral in construction, containing lime in quantity, and so frail that they are easily damaged.

The seaweeds, such as the Wracks, are essential to the shore, coast and estuary, because they provide a covering shelter from the waves, sun and wind for large numbers of marine creatures and also a food supply. Shores which have little or no *Algae* species are deficient in such marine creatures, too. One example is regularly moved shingle and pebbles, which does not allow the attachment and development of *Algae* spores.

The wave action, slope and shape of rocks, the temperature and saltness or otherwise of the water, light strength, all play a part in affecting the profic or sluggish growth of the *Algae,* but the British coastline is fortunate in having an abundance of several hundred species.

It has been possible to give only a general colour reference, because the seaweeds do change in colour due to circumstances. Reds have a tendency to browns and some browns are more olive or greenish, particularly during decomposition after being washed ashore.

Seaweeds have been useful to man for centuries. One early use was to dry and burn brown seaweeds, such as the Oarweeds, to obtain an ash known as "kelp". From this potash and soda were extracted for use in glass and soap manufacture. It is safe to say you have eaten or used seaweed in some way, though it is unlikely you were aware of this at the time. For example, extracts, usually alginic acid from the brown seaweeds, have been or are still used in icecream, jam, typewriter rollers, ink, bath foam, face cream, beauty soap, shaving cream, toothpaste, varnish, paint, plastics, cattle food, various fabrics, fireproof materials, plant fertiliser and slimming tablets and pills, to name a few. Seaweed extract helps jellies to set quicker and beer to clarify.

Seaweed contains iodine and was once a valuable source of this. Some species have been found to contain 22 important minerals, protein and various vitamins. As a food we lag behind the Chinese, Japanese and Hawaiians, the world's largest seaweed eaters, but in England, Wales and Ireland the Lavers are boiled to a jelly, which is then made into oatmeal-rolled "cakes" and fried, or stewed in milk by the Hebridean islanders and served with vinegar, butter and salt like spinach. Dulse is eaten raw or used as a cooking vegetable in Scotland. In Ireland Carrageen Moss is similarly boiled and eaten as a jelly or mixed with tea or coffee.

Research is still continuing to find more industrial and commercial uses for the abundance of seaweed on our coasts. Remember, however, that this prolific growth may not continue. Pollution causes seaweed growth to decline and it is already known that in the last century half the seaweed species on one stretch of English coastline have disappeared.

And without the seaweeds so, too, vanish the marine creatures and larger species dependent upon them.

GREEN SEAWEEDS

The Green Seaweeds (*Chlorophyceae*) occur most numerously on the upper parts of the shore, in the region below where the land plants reach the shore. An exception is where a stream or trickle of draining fresh water flows down the rocks or crosses the shore, particularly in estuaries. Here, along its course, some of the green seaweeds will be flourishing, although the middle and lower shore in the vicinity will probably be supporting brown and red seaweeds. Some of the green seaweeds will also be found in pools with a low salinity on the upper shore. In these seaweeds the chlorophyll or vital green pigment, is more dominant, than in brown or red seaweeds, where the chlorophyll is hidden by different pigments.

BRYOPSIS PLUMOSA

Common. It frequents the middle shore, where it occurs on the sides of rocks, also in deep pools. The plant's weak stems have lateral branches that are wider at the bottom and reduce in size progressing to the top, the branches having many branched threads, the latter creating a delicate feather-plumed or fernlike shape. In colour it is pale green and measures up to 4 inches in height.

Fig. 99 A Green Seaweed *Cladophora rupestris*

CLADOPHORA RUPESTRIS Fig. 99
Common. It frequents the middle shore where it occurs on shady rocks, but prefers less exposed positions and is often found mingled with other seaweeds, notably the Wracks. It can also be found in semi-fresh water or shallow, partially saline pools. The several, opposite, slender branches of thick "threads" give the plant an upright, tufted, moss-like appearance. It is dark green and measures up to 6 inches in height.

ENTEROMORPHA INTESTINALIS Fig. 100 (b)
Very common. It frequents most types of shore but more so in the splash zone pools, especially in areas where there is also fresh water on the upper shore. It is found, too, in estuaries and brackish water, attached to stones, rocks, shells, and similar objects. The long, irregular, tubular fronds, "nipped in" or compressed along their length, swell up with gas during growth. If torn away at this stage they float in the pool or sea, but if cast ashore they deflate and dry out white. Usually bright green in colour. It measures up to 24 inches in length and about $\frac{1}{3}$ inch in diameter.

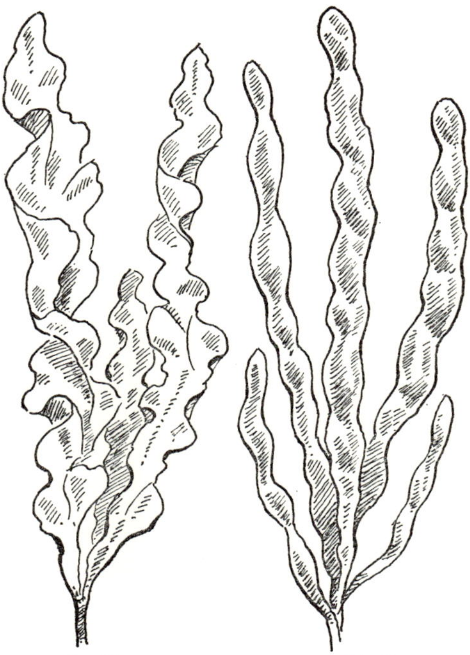

Fig. 100 (a) *Enteromorpha linza* (b) *Enteromorpha intestinalis*

ENTEROMORPHA LINZA Fig. 100 (b)
Common. It frequents most types of shore, in pools, on rocks, stones and other seaweeds and also occurs in areas where running fresh water joins the shore or pool. The single, elongated, narrow, flat frond is wavy or twisted spirally, while the stalk is hollow. It is bright green and measures up to 20 inches.

SEA LETTUCE OR GREEN LAVER (*Ulva lactuca*) Fig. 101
Common. It frequents the middle and lower shore, where it occurs in pools, both in salt or semi-fresh water, sometimes in groups or bunches, on rocks and stones, attached by a short stalk, but very frequently it is torn away and cast up on shores, or it may float on the sea in considerable masses. The large, broad, flat, semi-transparent frond is frail, very thin and membraneous with no definite shape. It is often an untidy mass on the shore, but in water has a wavy beauty. Because of a supposed resemblance to lettuce leaves it was given its name. It is edible and similar to the Purple Laver. In colour it varies from light green when young to bright or dark green according to its age. It measures up to 18 inches, although about 6 inches is more usual.

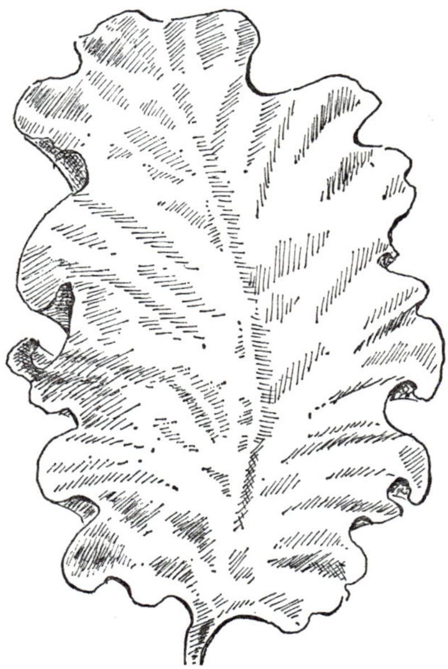

Fig. 101 Sea Lettuce or Green Laver *Ulva lactuca*

BROWN SEAWEEDS

The Brown Seaweeds (*Phaeophyceae*) include the extraordinary long Oarweeds and the curious jelly-like seaweeds, such as *Leathesia difformis*. The description brown is not strictly true in the normal sense of defining this colour as these seaweeds vary from a dark brownish to an olive-greenish; also if they become exposed and dry they may be almost black. The Wracks, in particular, especially those with bladders enabling them to float and obtain the maximum sunshine, are familiar on rocky coasts or when cast ashore, perhaps after a storm, from the zone where they usually occur. The brown seaweeds are either attached to rocks, live in deep water or are found in rock pools.

BLADDER WRACK (*Fucus vesiculosus*)

Very common. It frequents the middle shore where it is found in masses on rocks; it also occurs on seawalls of estuaries, pier piles, groynes, breakwaters, and similar objects, in brackish water, salt marshes, creeks and inlets. It is a tough, strong, leathery seaweed, with strap-like, flattish, lobed fronds, probably wavy-edged with conspicuous midribs, on each side of which there are also usually pairs of variously-sized, gas-filled bladders or swellings, hence its name. Because of the buoyancy the bladders give the seaweed in water it is able to resist heavier buffetings from the waves than other Wracks. It is olive-brown in colour and grows up to 3 feet in length.

BOOTLACE WEED OR SEA LACE (*Chorda filum*) Fig. 102 (a)

Very common. It frequents the lower shore, where it occurs adhering by a disc holdfast to stones, rocks, shells, and other objects. Sometimes it is torn away and cast ashore in rolled up bundles. The round, smooth, tough, slender, fronds are undivided and slimy, hence its name. It is brown in colour and measures about $\frac{1}{4}$ inch in diameter but up to 20 feet in length.

CHANNEL WRACK (*Pelvetia canaliculata*) Fig. 103 (a)

Very common. It frequents the upper part of the shore, even into the splash zone, occurring on rocks and may even be living close to coastal land plants in this habitat. The narrow, much-branched fronds have indented swollen tips and are marginally curled or inrolled to create a channel on one side facing the rocks. The purpose of this is to help the seaweed conserve moisture as it hangs downwards. As a result it can endure more exposure than other Wracks, but oddly enough is killed if covered by sea water for a lengthy time. It is olive-brown in colour

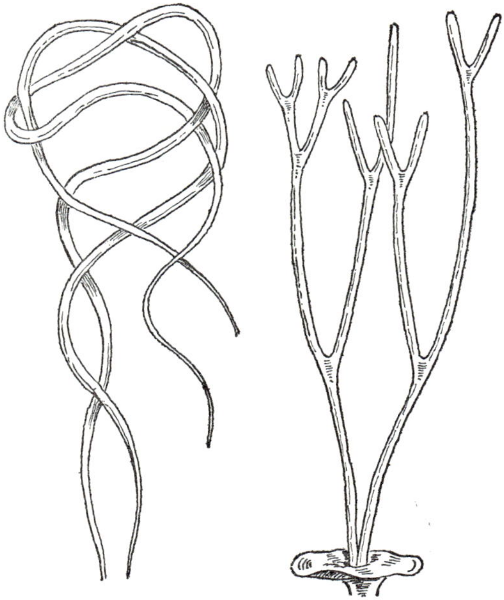

Fig. 102 (a) Bootlace Weed *Chorda filum* (b) Thong Weed *Himanthalia elongata*

and grows, in bunches, up to only 6 inches in length due to its exposed habitat.

FLAT WRACK (*Fucus spiralis*) Fig. 103 (b)
Common. It frequents the upper to the middle shore, but below the habitat level of the Channel Wrack, where it occurs on rocks. The smooth-edged, broad, twisted fronds have a prominent midrib and are similar to Bladder Wrack by being flattish, but do not bear the gas-filled bladders, although the blunt frond ends may be swollen with the fruiting bodies. It is olive-brown in colour and may measure up to 4 inches in length.

FUCUS CERANOIDES Fig. 103 (c)
Common. It frequents the middle shore where it occurs on rocks and stones; it is also common in brackish water, in estuaries, creeks, inlets, bays, at the mouth of streams and rivers where fresh water meets the shore. The noticeably thin smooth-edged fronds have a prominent mid-

rib and the ends also have a clearly-defined, sometimes swollen, pointed fork, but the bladders are absent. It is olive-brown in colour and may measure up to 3 feet.

KNOTTED OR EGG WRACK (*Ascophyllum nodosum*) Fig. 103 (d)
Very common. It frequents the middle shore but is more likely to be found in masses on sheltered beaches, particularly in estuaries, where it occurs attached to rocks and stones, as it cannot withstand the action of violent waves like other Wracks and is soon torn away or damaged. It is easier to identify because the long, irregularly-branched fronds do not have a midrib but, spaced singly along the middle of the frond lengths, are usually wider, variously-sized oval or egg-like bladders, while the yellowish fruiting bodies appear, sometimes in pairs, from the sides of the fronds. Because of this growth it has a rather untidy appearance. In some areas it is the dominant middle shore seaweed to the exclusion of any others. It is dark olive-green and may measure up to 4 feet in length, sometimes longer.

LEATHESIA DIFFORMIS
Very common. It frequents the middle and lower shore where it occurs on other seaweeds and rocks. The rounded, lobed frond does not look at all like what is normally thought of as seaweed and has the appearance of a jelly-like lump attached to another seaweed. It is more likely to be found in the spring or summer months. It is olive-brown and measures up to 2 inches across.

OARWEED OR TANGLEWEED (*Laminaria digitata*) Fig. 104 (a
Very common. It frequents the lower shore where it occurs on rock and is sometimes so abundant that it forms a thick "carpet" when th tide is out. It is also found in large rock pools. The large, tough rubbery, wide frond is usually divided into "fringes", while the long flexible, smooth stalk, with an oval cross-section, is attached by a cluste of intertwining, thin, flat, spreading, branching "rootlets" to form strong holdfast which clings to the rock. Even so, during a storm th Oarweeds are frequently torn from their habitat and the fronds washe ashore. It is brown in colour and may measure up to 6 feet in lengt and 2 feet across. A similar common Oarweed is *Laminaria hyperbore* occurring in the same habitat, but having a rough, rigid stalk, with round cross-section; also the "rootlets" forming the holdfast are thic and short, bending like fingers gripping the rock.

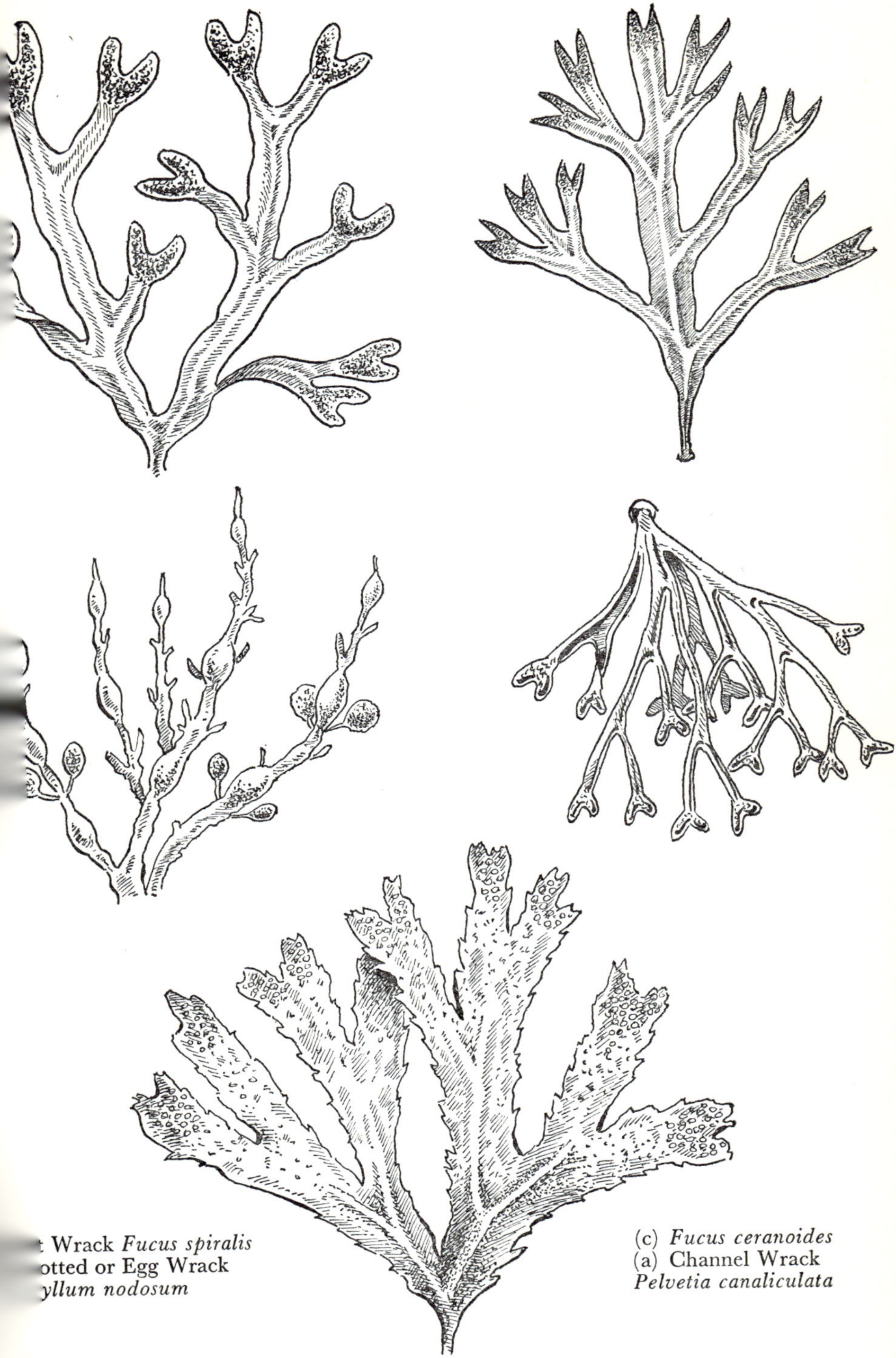

t Wrack *Fucus spiralis*
otted or Egg Wrack
yllum nodosum*

(c) *Fucus ceranoides*
(a) Channel Wrack
Pelvetia canaliculata

(e) Serrated or Saw Wrack *Fucus serratus*

Fig. 103 Some Seaweeds

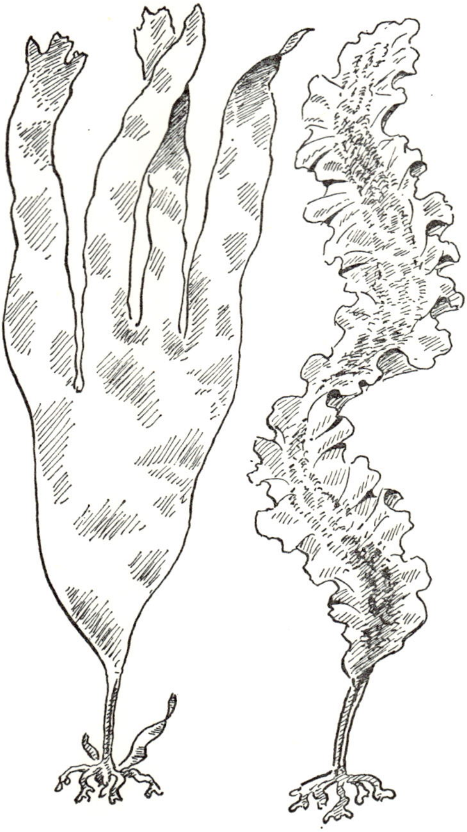

Fig. 104 (a) Oarweed *Laminaria digitata* (b) Oarweed *Laminaria saccharina*

OARWEED OR TANGLEWEED (*Laminaria saccharina*)
Fig. 104 (b)

Very common. It frequents the middle and lower shore, where it occur on stones and rocks; it is also found in large pools. After storms it i often washed ashore on flat sandy or muddy sand shores, such as bays and the fronds are a familiar item among debris to anyone beachcomb ing close to the edge of the sea. The Latin name was given to it due t its containing a type of sugar, mannitol. An aid to identification fron the previous two species is that it has an undivided, flat, wavy-edged wrinkled, rubbery frond with a shorter, smooth round stalk, but n midrib. It is sandy-yellowish colour and usually measures up to 9 feet in length, but sometimes more, and up to 12 inches across.

SERRATED OR SAW WRACK (*Fucus serratus*) Fig. 103 (e)
Very common. It frequents the lower middle shore and lower shore where it occurs on rocks. As its name describes, the flattish, irregularly branched fronds, with a prominent midrib, have serrated or toothed edges, but do not have the gas-filled bladders. It is brown in colour and measures up to 6 feet in length.

THONG WEED (*Himanthalia elongata*) Fig. 102 (b)
Locally common. It frequents the lower shore where it occurs on rocks. The frond is a concave, stalked, mushroom-like growth, which develops from a round, button-like knob. Out of the centre of the frond grows a leathery, several-times branched strap, which bears the fruiting bodies. Because of this construction it was so-named and is very easy to identify. Sometimes this mushroom-like base is still attached to the habitat after the strap has been torn away. It is olive-brown in colour. The frond is up to 2 inches in height, while the strap may be up to 6 feet in length, occasionally longer.

RED SEAWEEDS

The so-called Red Seaweeds (*Rhodophyceae*) are among some of our most beautiful and delicate seaweeds, varying widely in form, but are seen less frequently than the Brown and Green Seaweeds. "Red" has to be literally interpreted as a shade from blood red, pink, to a purplish hue, but I have given the most approximate colour possible.

CARRAGHEEN MOSS OR IRISH MOSS (*Chondrus crispus*) Fig. 105
Very common. It frequents the middle shore in varying rocky habitats xcept muddy areas. In rock pools it is often at its colourful best. The ough, flat fronds are wide and much divided into branches that give n overall "rounded" shape to the seaweed, but its shape is conditioned y the habitat. In a sheltered pool this edible seaweed is numerously ranched, almost stalkless and grows prolifically to carpet the flat rocks, ut on an exposed shore, where the rock angles may not be entirely itable, it will have a stalk and the fronds are narrower, with few ranches. It is also extremely variable in colour and may be a dull dark d, purplish-red, rose red, violet red, pinkish, brownish-yellow, even

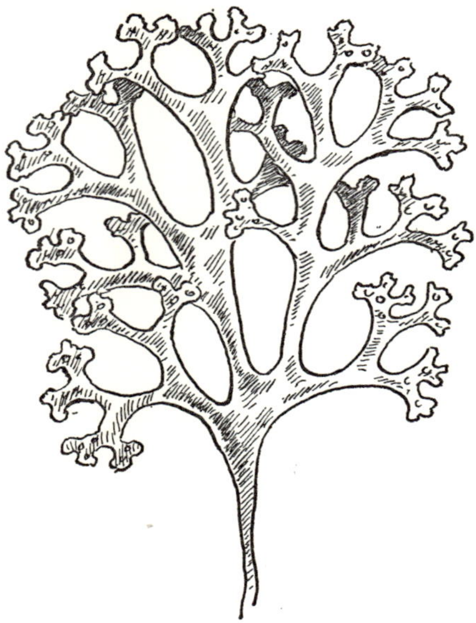

Fig. 105 Carragheen Moss or Irish Moss *Chondrus crispus*

pale yellowish or green. In the water it has a beautiful irridescent sheen which disappears as soon as it is removed from the pool. Sometimes it may be found incorporating several of these colours. It grows up to 6 inches in length.

CERAMIUM RUBRUM

Very common. It frequents the middle shore where it occurs in tidal pools and on rocks and stones, sometimes also attached to other sea-weeds. The plant is variably much branched, the banded, filamentous threads branching in pairs, and grows in tufts on the habitat. It varies in colour, from dark red to a reddish-brown and measures up to 10 inches in length.

CORALLINE OR CORAL-WEED (*Corallina officinalis*) Fig. 106

Very common. It frequents the middle shore where it occurs in masses on the edges of rock pools and is at its best in the shade or semi-shade. It is easily recognisable because it grows upright and has a hard, rough encrusting limy coat. The bushy tufts of thin, opposite stems and divided stiff branches comprise small "joints" or articulated sections so that the plant is also pliant. In the past naturalists were undecided if Corallina

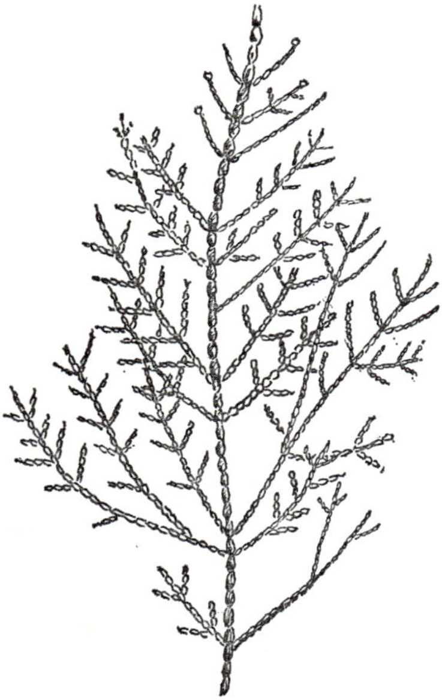

Fig. 106 Coralline *Corallina officinalis*

Fig. 107 Delesseria *Delesseria sanguinea*

was a plant or animal, due to its being able to extract lime and create this hard coating, in the same way as polyps and corals thus it was so-named, but when the "skeleton" is dissolved a vegetable construction is revealed proving it belonged among the seaweeds. It is usually a deep purple or purple-red to yellowish. After it dies and decays the limy "skeleton" remains and dries out white. It measures up to 6 inches in height.

DELESSERIA (*Delesseria sanguinea*) Fig. 107
Common. It frequents the lower shore where it occurs in the larger, deeper, shaded pools, often as a separate plant. This species is quickly recognisable by having a short, several branched stem with large, elongated, smooth-edged, wavy, delicate, leaf-like fronds, each having a conspicuous midrib and "veins", so that they similarly resemble in shape the leaves of the Sweet Chestnut tree, but are not so stiff. They are an attractive, bright pink or deep red colour. The plant measures up to 6 inches in length.

DULSE, DILLOSK OR DILLISK (*Rhodymenia palmata*)
Very common. It frequents the middle and lower shore where it occurs on rocks, also upon seaweeds, such as the Wracks and Oarweed. The tough, irregular, indented fronds have been likened to the shape of a hand and its fingers, but it varies considerably from palmated to fan-like or a single, wide, flat frond with varying-sized blades protruding from the sides. It may be growing thickly in tufts on the rocks or just a few plants forming a fringe at the edge of the pool. In colour it is a deep red or reddish-brown, with purplish tints upon the fronds when in clear water. It is edible and measures up to 12 inches in length.

LOMENTARIA ARTICULATA
Very common. It frequents the lower shore where it occurs in pools, also upon other seaweeds, rocks and stones. The plant is easily recognised because the irregular, divided, jointed branches are "nipped" or compressed at intervals along their length so that they have the appearance of being made of a child's oval beads strung together. It is reddish-pink in colour and measures up to 8 inches in length.

MEMBRANOPTERA ALATA
Very common. It frequents the middle and lower shore where it occurs in pools, on rocks and among and upon other seaweeds, preferring a shady situation to one in direct sunlight. The very flat, narrow fronds are ribbon-like, with a prominent midrib, and divided into midribbed branches with forked tips. It is blood-red, sometimes pinkish, and measures up to 8 inches in length.

PLUMARIA ELEGANS Fig. 108
Very common. It frequents the middle and lower shore where it is limply suspended from overhanging rock ledges, and also prefers a shady situation. The whole plant has a frail delicacy because the small, soft, feathery, opposite branches are much divided and sub-divided. It is a deep crimson-red colour and up to 4 inches in length.

PURPLE LAVER OR SLOKE (*Porphyra umbilicalis*)
Very common. It frequents all levels of shores, particularly on rocks and large stones in exposed, sandy areas. The broad, flat frond is thin and membraneous with no definite shape except for a wavy-lobed outline, seen when the plant is spread out. It may be found undamaged and attached to the habitat, but is easily torn away and the complete

Fig. 108 *Plumaria elegans*

plants or portions of it are cast up in a heap on the sandy shore. It is edible, like the similar Green Laver or Sea Lettuce. In colour it is shades of purple, usually a purplish red or purplish-brown, and measures up to 10 inches in length.

Chapter 16

Flowering Plants and Grasses

More than 5000 plants have been recorded growing in the British Isles. Of this flora a large number occur in the habitats covered by this book's title. There are also many common inland plants that are equally at home within sight and sound of the sea. The visitor to the coast, estuary and seashore will be able to find the familiar buttercups, herb robert, kidney vetch, ragwort, sheep's bit, small scabious, scarlet pimpernel, spotted medick, hop trefoil, strawberry clover, goosegrass, common mallow, furze, the brambles, common yellow toadflax, ox-eye daisy, to name only a few which elsewhere live in sometimes very different types of habitat.

There are, however, a considerable number of wild plants which occur only on or by the coast, estuary or seashore. Like the maritime grasses I refer to later, these plants enjoy the specialised conditions here, be they the salty air or spray-laden breezes, the heavily saline soil, others finding the arid conditions to their liking, while some occur close to brackish water. But very few flowering plants thrive by complete submergence in the sea. Those that do I have described.

Some species can be found over a varying extensive area, for example from the salt-marsh to the drift line perhaps low on the foreshore, but the observer will also note that others are more restricted and stay in a particular habitat. If the conditions are different even only yards or feet away the plant will not extend away from the habitat it can exist on. To maintain themselves securely where the habitat is loose or unstable some species produce an extensive, deeply penetrating root-stock. For example, when growing in shingle or clinging to the fissures in the face of an exposed chalk cliff. Many of the coast and seashore plants are glaucous, their stems and leaves being covered with a layer of wax, giving them a bluish-grey or green appearance. The wax is essential to survival in a dry habitat, because it prevents unnecessary loss of the water which the plant has been able to store. Like inland plants those in the described habitats reproduce by seed or extend vegetatively by creeping shoots on the surface or underground stems which project new shoots periodically.

To indicate the likelihood or otherwise of finding a plant I have used simple definitions. "Common" means just that, the species is frequently to be found on the coast, estuary or seashore, in the particular type of occurring habitat mentioned. If it also occurs inland this is stated. "Locally common" means that a plant may be found in large numbers, or frequently, in one type of habitat, but is not found in the same numbers or at all, elsewhere. "Uncommon" means a scarce plant is found only as a few examples in a habitat or growing separately and scattered over a large area. "Rare" describes a plant that is found infrequently, perhaps occurring as a single example in a habitat.

ALEXANDERS (*Smyrnium olusatrum*) [*Carrot Family-Umbelliferae*]
Locally common, sometimes abundant. It occurs on waste ground and in hedges, on chalk cliffs, near the sea. It has an upright, round, stout, ribbed stem bearing several compound umbels of densely massed greenish-yellow flowers. The large glossy green, toothed, lower leaves are divided three times into oval leaflets, the stalks being extended into wide sheaths at their base, while the smaller, opposite, oval, three-part stem leaves also have broad, hairy sheaths on their main leaf-stalks, which is an aid to identification. As the shoots and stalks were boiled as a pot-herb and eaten like celery in the past it may originally have been an escape from cultivation. It grows up to 4 feet in height. Biennial. Flowering period is April to June.

BLACKTHORN OR SLOE (*Prunus spinosa*) [*Rose Family—Rosaceae*]
Common. It occurs on cliffs and in waste places, hedges, near the sea, sometimes creating a dense, near-impenetrable thicket. A frequently-branched shrub with dark brownish-black or dark grey bark on twigs and stems which also bear numerous tough thorns. It may reach up to 10 feet in height. The white flowers, like the related plum, appear in pairs or singly before the oval or elliptical leaves open. The blue-black stone fruits are small, almost round and have a bluish "bloom" upon them, but be wary of eating them unless they are very ripe as the "tartness" will set your teeth "on edge" for a considerable time. Formerly they were much sought to make the esteemed sloe gin and a jam, and

the wood being used when toughness was essential. Deciduous. Flowering period March to May.

BLOODY CRANE'S-BILL (*Geranium sanguineum*) [*Geranium Family—Geraniaceae*]
Locally common. It occurs in dry, grassy areas, in sandy places, on dunes, among rocks, on cliff ledges, particularly limestone and chalk. The numerous, hairy, slender stems, bent at the joints, measure up to 24 inches in length and may be up to 12 inches in height, and bear roundish leaves deeply cut into several segments, each segment also usually being notched. The attractive, round, magenta or purplish-crimson flowers are about 1½ inches in diameter and separate on long stalks. Seed pods are long and pointed with a supposed resemblance to a Crane's bill. Perennial. Flowering period June to August.

BLUE FLEABANE (*Erigeron acris*) [*Daisy Family—Compositae*]
Locally common. It occurs on dunes, stable shingle and dry grassy areas, also inland. The stiff, upright, sometimes crimson stem, up to 18 inches in height, but less in dry habitats, bears numerous short, lance-shaped leaves, which decrease in size up the stem, while the base leaves are broader and tapering. The hairy stem bears several branches of composite flowers, half an inch or less across, the disk florets being yellow and the ray florets dull purplish, despite its name. Annual or biennial. Flowering period July to September.

BRISTLY OX-TONGUE (*Picris echioides*) [*Daisy Family—Compositae*]
Common. It occurs in waste, grassy places near the sea coast, estuaries and salt marshes, also inland. The branched stem is up to 3 feet in height, the whole plant being very bristly with hooked hairs. The stalkless, lance or heart-shaped leaves also bear large bristles from raised white "pimples". At its base the rosette of stalked root leaves, which lie flat upon the soil, are broad lance-shaped. The composite light yellow flower-heads appear in branched clusters. If the stem is broken a milky juice exudes. The brownish achenes have a long slender "beak" with a plumose "parachute", similar to the dandelion, for dispersal. Annual or biennial. Flowering period June to September or October.

BUCK'S HORN PLANTAIN (*Plantago coronopus*) [*Plantain Family—Plantaginacea*]
Locally common. It occurs on bare, dry, sandy, stony and shingly

places near the sea coast, also inland. A small, downy plant, up to 6 inches in height, the narrow linear, sometimes deeply toothed or divided lobed, hairy leaves form a rosette. The green flowers with yellow stamens form short spikes, about 1½ inches in length, on separate, slender stalks. See SEA PLANTAIN. Biennial. Flowering period May to August.

COMMON CENTAURY (*Centaurium erythraea*) [*Gentian Family—Gentianaceae*]
Common. It occurs in dunes, stable shingle, and in dry pastures, particularly on the turf of chalk and limestone cliffs. The square stem measures up to 12 inches in height, depending on habitat, bearing a few, opposite pairs of narrow oval leaves, which join at their bases, the upper leaves being smaller, there also being a rosette of leaves at stem's base. The delicate pink, numerous flowers are funnel-shaped, blooming in crowded tufts. Annual. Flowering period from June to September. A related species is the SLENDER CENTAURY (*Centaurium pulchellum*) which is locally common in salt marshes and dry places near the sea. As its name indicates its stem is shorter and more slender and usually has several flowering branches from its base, which bear broader, egg-shaped leaves, but there is no rosette of leaves around the base. The smaller flowers are darker, more reddish than pinkish and bloom further apart. Flowering period July to October. Another species, the SEASIDE CENTAURY (*Centaurium littorale*) is shorter, but has larger, bluish-pink flowers and thick, narrow, strap-shaped leaves. It is uncommon, occurring on sand dunes and dry places near the sea in north-west England and south-west Scotland. Flowering period July.

COMMON OR LESSER DODDER (*Cuscuta epithymum*) [*Bindweed Family—Convolvulaceae*]
Locally common. It occurs on stable plant-colonised shingle, also downland and cliff slopes, where furze, heath, thyme and other low shrubs and plants grow. It is a parasite, feeding on its host by means of penetrating disc-like suckers, its leafless, thread-thin, crimson stems climbing in an anti-clockwise, close twining tangle around its host's stems. The very small, waxy, bell-like red and white stalkless flowers bloom in globular bunches. The seeds which result do not germinate until late in the following year when there is again sufficient host vegetation to parasiticise. Annual. Flowering period July to September.

COMMON MELILOT (*Melilotus officinalis*) [*Pea Family—Papilionaceae*]

Common. It occurs in waste places, salt marshes and stable shingle near the coast, estuaries and mudflats. A tall plant, up to 4 feet in height, the erect, branched stem bears numerous lengthy spikes of small, light yellow, pea-like flowers and narrow, long-stalked trefoil leaves. The smooth seed pods are brown or olive. Annual or biennial. Flowering period June to August. Related is the WHITE MELILOT (*Melilotus alba*), an alien introduction, which is similar to the COMMON, but has white flowers. It is taller and occurs in waste places, sandy areas and salt marshes close to the shore, in estuaries and near mudflats. Less common than the COMMON but increasing in frequency and range. Biennial. Flowering period July to September.

COMMON ORACHE OR SPREADING ORACHE (*Atriplex patula*) [*Goosefoot Family—Chenopodiaceae*]

Very common. It occurs in waste places by the sea, also inland as a pest. The prostrate or upright, several branched, mealy plant, up to 3 feet in height, has reddish stripe lines on the stem, which are an aid to identification. The opposite leaves are variable, toothed, lance-shaped, with a wedge-like base, or triangular, the upper leaves being narrow. The petal-less, tiny green flowers grow in clusters on spikes. Annual. Flowering period July to October. Related is the locally common, similar FROSTED ORACHE (*Atriplex laciniata*) occurring on sandy seashores and coastal waste places, an aid to identification being the silver-white hue of the broad, triangular, wavy-edged, toothed leaves and buffish-brown or reddish, spreading stems. Annual. Flowering period the same as COMMON ORACHE. The GRASS-LEAVED OR NARROW ORACHE (*Atriplex littoralis*) also commonly occurs in waste places near the sea, up to 2 feet in height, with erect branches bearing linear leaves, hence its name. The flowers are similar to COMMON ORACHE. Annual. Flowering period July to September.

COMMON OR ANNUAL SEABLITE (*Suaeda maritima*) [*Goosefoot Family—Chenopodiaceae*]

Common. It occurs on salt marshes and sandy and muddy shores c coast and estuaries. It measures up to 12 inches in height when erect or may be semi-erect or trailing prostrate, the branched stems bearin numerous, succulent, semi-cylindrical, linear, pointed, bluish-green c reddish-tinged leaves. The tiny green flowers are almost hidden amon the leaves. Annual. Flowering period July to October. Related is th uncommon SHRUBBY SEABLITE (*Suaeda fruticosa*), which occurs on stable shingle shores and sand, also salt marshes, chiefly on the south and east coasts of England. It is shrub-like, usually up to 3 feet in height, sometimes more, with thick, succulent stems and much branched, bearing densely numerous, alternate, round-tipped, thick, linear, glaucous-green leaves. The flowers, with the same flowering period, are similar to the COMMON SEABLITE. Perennial.

EEL-GRASS, SEA GRASS OR GRASS WRACK (*Zostera marina*) [*Eel-Grass Family—Zosteraceae*] Fig. 109
Uncommon. Formerly common, then disappeared from many previous habitats, but now showing signs of recovery. It occurs in estuaries, creeks and similar where the coast is protected, on the middle and lower shore and below, on mud, muddy sand and gravel, where it creates beds of what appears, when exposed, to be grass or grass-like green seaweed. These are the only British *flowering* plants that can actually live and grow, with creeping roots, submerged by the sea. The flat, several-

Fig. 109 Eel-Grass or Sea Grass *Zostera marina*

veined, dark green, narrow, alternate leaves measure up to 3 feet in length and ¼ inch in width. The petal-less flowers, enclosed in a leaf-like sheath, appear in spikes at the base of some of the leaf branches, the water currents aiding fertilisation of the flowers. Perennial. Flowering period June to August. The related, uncommon, NARROW-LEAVED EEL-GRASS (*Zostera hornemanniana*) is smaller, up to 12 inches in length, the leaves being similar but narrower, with up to three veins. It occurs usually in muddier habitats. Perennial. The uncommon DWARF EEL-GRASS (*Zostera nana*) is even smaller, up to 6 inches in length, the one-veined leaves being extremely narrow, nearly mere threads, with a notch at their tip. It occurs on the middle shore, more likely on mud or muddy sand. Perennial.

ENGLISH STONECROP (*Sedum anglicum*) [*Stonecrop Family—Crassulaceae*]
Locally common. It occurs on stable, sandy shingle, dunes, stony places, cliff banks, rocks, near the sea, where it sometimes covers an extensive area with its spreading mat-like tufts. Despite its name it occurs also in Wales, Scotland and Ireland. A little glabrous, succulent plant, up to 3 inches in height, its short, branching and creeping stems bear numerous small, thick, egg-shaped, water-storing, glaucous green or pinkish leaves. The starlike, pink-tinged, white flowers are clustered to form the heads or cymes. Perennial. Flowering period May to August.

FENNEL (*Foeniculum vulgare*) [*Carrot Family—Umbelliferae*]
Locally common. It occurs on dry waste places and cliff faces near the sea, also inland. The erect, jointed, round, grooved, tubular, branched stem measures up to 4 feet in height and bears leaves which are s deeply divided that they form soft, very narrow threadlike, bluish-gree segments, the leaves having large sheathlike bases where they join th stem. The tiny yellow flowers are clustered together to form large com pound heads or umbels. It has a rather pungent aromatic odour. I the past it was commonly used as a pot herb and in a sauce for fish, fo which it was formerly widely cultivated. Perennial. Flowering perio July to September.

GLASSWORT OR MARSH SAMPHIRE (*Salicornia europea*) [*Goosefoot Family—Chenopodiaceae*] Fig. 110
Common. It occurs on areas of salt marshes, mudflats and mudd seashores which are submerged at high tide. An aid to identificatio is that it has no leaves and the flowers are inconspicuous. The ere

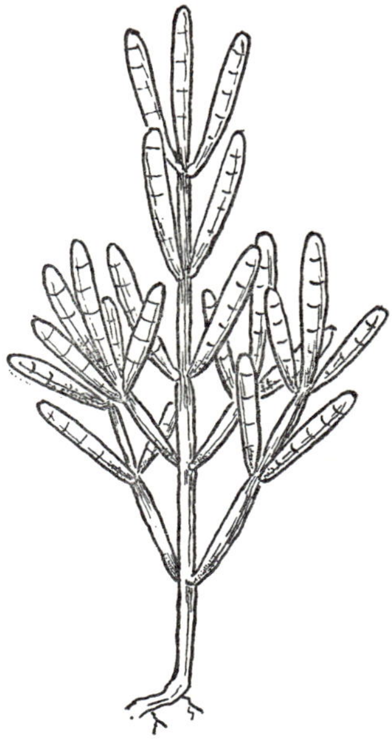

Fig. 110 Glasswort *Salicornia europea*

succulent, yellow-green or bright green, cylindrical, jointed stem, bearing pairs of side branches which are themselves usually branched, measures up to 12 inches in height, occasionally more. The tiny, petalless green flowers are produced in threes at or near the junctions of the jointed branches. At first glance it has the appearance of a succulent green cactus, but after summer changes to red. Annual. Flowering period July to September. There are several related Glassworts. The Locally common CREEPING OR PERENNIAL GLASSWORT (*Salicornia perennis*) has a creeping rootstock which produces tufts of succulent stems, either brownish, crimson or greenish-orange. It occurs on firm mud and stony salt marshes, chiefly in Southern England, Uncommon elsewhere. Flowering period August to October. The locally common BUSHY GLASSWORT (*Salicornia dolichostachya*) develops as a loose round, branched, green bush on soft mud. Locally common SEABLITE GLASSWORT (*Salicornia prostrata*) is a red, prostrate plant occurring on firm mudflats in Southern England. The rarer FRAGILE GLASSWORT (*Salicornia pusilla*) has solitary flowers and grows on firm mud in Southern England. The common TWIGGY GLASSWORT (*Salicornia ramosissima*) occurs on firm mud; the smaller, green, yellow or purplish-red stem segments give the plant a bead-like appearance.

GOLDEN SAMPHIRE (*Inula crithmoides*) [*Daisy Family—Compositae*] Fig. 111 (b)

Locally common. It occurs on salt marshes, seawall banks and creeks, also among rocks on sea cliffs. The erect stem measures up to 18 inches in height, bearing numerous alternate, smooth, narrow linear, thick succulent, yellow-green leaves, some of which are three-pointed at their tip. The attractive, daisy-like bright golden-yellow flowers, each one inch wide, are produced in loose clusters. Perennial. Flowering period July and August.

Fig. 111 (a) Samphire *Crithmum maritimum* (b) Golden Samphire *Inula crithm*

HARE'S-FOOT TREFOIL OR CLOVER (*Trifolium arvense*) [*Pea Family—Papilionaceae*]

Common. It occurs in sandy places and on stable shingle and dune near the sea, also inland. Easily recognised from other species of Clove and Trefoils by the oblong, soft, downy heads, up to an inch in lengt of small, pinkish-white flowers, on long stalks. The erect, many-branche stem, clothed with short, soft hairs, up to 12 inches in height, bears narrow, trefoil leaves. Annual. Flowering period July to September.

KNOTGRASS (*Polygonum aviculare*) [*Dock Family—Polygonaceae*]
Common. It occurs on banks and waste places near the sea, in estuaries, creeks and salt marshes, also inland. Despite the confusing name it is not a Grass, being a semi-prostrate, much-branched, straggling plant, the wiry slender stems, 12 inches or more in length, bearing short, lance-shaped, but variable, alternate leaves which have frayed, silver-white or brownish sheaths at their base. The small white or pink, occasionally pale green, flowers appear in the axils of the leaves on the stem and create the "knotted" resemblance. Annual. Flowering period May to October. This species should not be confused with the rare SEA KNOTGRASS (*Polygonum maritimum*) which is similar, occurring on sandy shores, but has much thicker, larger, grey leaves, with their edges turned down, their undersides being glaucous, the silver sheaths and leaves also having branched "veins". Perennial. The larger flowers appear July to September. The SLENDER SEA KNOTGRASS (*Polygonum raii*) is also uncommon, similar to the Knotgrass, but has larger, shiny red-brown and thicker stems, occurring on shingle and sandy shores. Flowering period August to September.

LARGE EVENING PRIMROSE (*Oenothera erythrosepala*)
[*Willowherb Family—Onagraceae*]
Locally common. It occurs in sandy, stony dunes and waste places near the sea, also salt marshes and creeks and inland. In some places it is almost certainly a garden escape that has established itself. It grows up to 4 feet in height, the stiff, hairy stem bearing broad, lance-shaped leaves and spikes of large, pale yellow flowers. Annual or biennial. As its name indicates the flowers open in the evening. Flowering period June to September. Similar is the locally common FRAGRANT EVENING PRIMROSE (*Oenothera stricta*) but it is smaller than the LARGE with correspondingly narrower leaves and stem, the pale yellow flowers being noticeably fragrant. It occurs in sandy areas. Flowering period June to September.

MARSH MALLOW (*Althae officinalis*) [*Mallow Family—Malvaceae*]
Locally common. It occurs on salt marshes and dykes close to the sea. The erect stem, up to 4 feet in height, bears large, alternate, nerved, heart-shaped, oval or rounded, greyish-green leaves, which are usually

three or five lobed. The delicate, pale rosy-pink flowers, an inch or more wide, appear in clusters from the base of the stem leaves. In shape they have a similarity to those of the hollyhock. An aid to identification is that the whole plant feels velvet-like when handled due to its dense clothing of soft hairs. Perennial. Flowering period August and September.

PRICKLY SALTWORT (*Salsola kali*) [*Goosefoot Family—Chenopodiaceae*] Fig. 112

Common. It occurs in salt marshes and on sandy shores. The stout, much branched, pink or green striped, thick stem may be prostrate or upright, up to 18 inches in length, bearing narrow, lance-shaped, thick, fleshy leaves, each of which has a sharp tough prickle at its tip, hence its name. The tiny stalkless greenish flowers appear singly, each with two sharp-tipped leaf-like bracts, at the base of the upper leaves. Annual. Flowering period July to September.

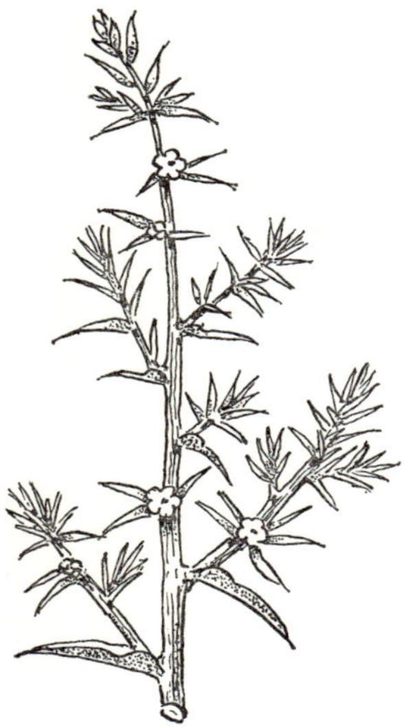

Fig. 112 Prickly Saltwort *Salsola kali*

RED SPUR-VALERIAN (*Centranthus ruber*) [*Valerian Family—Valerianaceae*]

Locally common. It occurs on sea cliffs, often in exposed positions, also on old walls and ruins near the sea. In some places it may originally have been a garden escape. The woody stem produces numerous tough smooth branches, up to 3 feet in height, bearing thick, stalkless, opposite, pointed, oval lance-shaped or egg-shaped greyish-green leaves. The numerous deep crimson red, pinkish or white flowers are clustered together to form blunt conical flower heads. Perennial. Flowering period May to September.

SAMPHIRE (*Crithmum maritimum*) [*Carrot Family—Umbelliferae*] Fig. 111 (a)
Locally common. It occurs on sea cliffs and rocky coasts, chiefly in south-east, southern and western England, sometimes growing in rock crevices on the upper shore near the high tide level. The stout solid stems, up to 12 inches in height, bear fleshy leaves much divided into numerous narrow leaflets, the tiny white-petalled flowers, with yellow ovaries, being grouped into compound umbels. An aid to identification is that all the plant is thick, succulent, and has a glaucous "bloom", with a striking contrast between the greyish-green foliage and yellowish flower umbels. In some habitats the young leaves are still gathered in spring for pickling, as they were frequently in the past. When bruised it has an aromatic fragrance. Perennial. Flowering period June to September.

SCURVY-GRASS (*Cochlearia officinalis*) [*Cabbage Family—Cruciferae*] Fig. 113 (d)
Common. It occurs on damp sea cliffs and muddy seashore banks. The several fleshy stems, up to 12 inches in height, bear fleshy, blunt lobed or toothed, stalkless, eared leaves that clasp the stem, while the long-stalked dark green root leaves at the plant's base are approximately heart-shaped, kidney-shaped or rounded. The clusters of four-petalled flowers are white or pale lilac. It is not related to the Grasses and derives its name from a former medicinal use to counter scurvy. Annual or Biennial. Flowering period May to August. The related common EARLY SCURVY-GRASS (*Cochlearia danica*) occurs on stable shingle, also muddy, rocky and sandy shores. It is usually prostrate and smaller, all the heart and ivy-shaped leaves being stalked, the white or lilac flowers blooming from March to September.

SEA ARROW-GRASS (*Triglochin maritima*) [*Arrow-grass Family—Juncaginaceae*]
Common. It occurs in salt marshes. The leafless flower stems measure up to 12 inches in height, the inconspicuous, small yellow-green flower spikes having a similarity to the SEA PLANTAIN. The radial leaves

Fig. 113 (a) Sea Aster *Aster tripolium* (b) Sea Campion *Silene maritima* (c) Sea Lavende *Limonium vulga*
(d) Scurvy Grass *Cochlearia officinalis* (e) Sea Plantain *Plantago maritima* (f) Thrift or Sea Pink *Armeria maritim*

are long, narrow, thick and rather succulent. It is not a Grass, the name arising from the arrow-shape of the ripe fruiting body of the related MARSH ARROW-GRASS (*Triglochin palustis*). Perennial. Flowering period May to September.

SEA ASTER (*Aster tripolium*) [*Daisy Family—Compositae*]
Fig. 113 (a)
Locally common. Occurs in salt marshes, edges of muddy creeks and mudflats, cliff edges and similar places by the sea. The erect, branched, stout stem, up to 3 feet in height, but more usually around 12 inches, bears smooth, succulent, oblong or lance-shaped, dark green leaves. The loose clusters of flowers are an aid to identification as they have a striking similarity to the garden michaelmas daisy, with tubular, yellow disk-florets circled by one whorl of purple-violet, bluish-mauve, lilac or, less often, whitish ray-florets, which may be absent. Perennial. Flowering period July to September.

SEA BEET (*Beta vulgaris*) [*Goosefoot Family—Chenopodiaceae*]
Fig. 114 (b)
Very common. It occurs in salt marshes, near the edges of mudflats and creeks, on seawall banks, rocky and muddy-sandy shores, the bases and ledges of cliffs, sometimes in the splash zone, and similar. It is a rather untidy plant the stems spreading in a floppy manner, but when erect they are up to 2 feet in height. They bear large, wavy, broad, triangular, thick, fleshy, shiny, dark green lower leaves, looking as if varnished, the upper leaves being smaller and narrower. The small green flowers are clustered in long, leafy spikes. This plant was the forerunner of the cultivated beetroot, mangel-wurzel and sugar beet. Formerly the root leaves were gathered, boiled and eaten as a substitute for spinach. Perennial. Flowering period June to October.

SEA BINDWEED OR SEASIDE CONVOLVULUS (*Calystegia oldanella*) [*Bindweed Family—Convolvulaceae*]
Uncommon. It occurs on sand dunes and sandy places, sometimes hingle, near the sea. The long, prostrate, creeping, slender stems, which do not usually twine like other related species, bear numerous, alternate, leshy, long-stalked, rounded or kidney-shaped, pale green leaves, with arge, white-striped, rosy-pink or pale mauve, funnel-shaped flowers, about 2 inches in diameter. Perennial. Flowering period June to August.

Fig. 114 (a) Wild or Sea Cabbage *Brassica oleracea* (b) Sea Beet *Beta vulgaris*

SEA-BUCKTHORN OR SALLOW-THORN (*Hippophaë rhamnoides*) [*Oleaster Family—Elaeagnaceae*]
Locally common. It occurs on cliffs, sand dunes and sea coasts, mainly in eastern England. A spiny shrub, measuring up to 6 feet in height sometimes more, bearing alternate, narrow, lance-shaped, silvery, scaly leaves, similar to a willow. The tiny green flowers appear prior to the leaves. The fruit is an orange or yellowish berry. Flowering period April and May.

SEA CAMPION (*Silene maritima*) [*Pink Family—Caryophyllaceae*] Fig. 113 (b)
Common. It occurs on stable shingle, rocky and sandy shores and cliff ledges. A prostrate spreading plant, the stems radiate, tinged purple up to 6 inches in height, bearing opposite, stalkless, narrow lance-shaped bright green leaves. The few scattered flowers on long stalks are white with five, deeply lobed, petals, the calyx being inflated, net-veined and green tinged with crimson. Similar to the BLADDER CAMPION

found inland in dry waste places. Perennial. Flowering period May to August.

SEA CLOVER (*Trifolium squamosum*) [*Pea Family—Papilionaceae*] Uncommon. It occurs in salt marshes and meadows near the sea, in southern England. The hairy prostrate stems, up to 6 inches in height, bear narrow trefoil leaves and oval-shaped heads of tiny light pink flowers, the latter being terminal above the last leaves. Annual. Flowering period June and July.

SEA HOLLY, SEA ERYNGO, SEA HOLM OR SEA HULVER (*Eryngium maritimum*) [*Carrot Family—Umbelliferae*] Fig. 115 Locally common. It occurs on sandy shores. This plant is immediately recognisable by its stiff much-branched spreading stems, up to 2 feet in height, of leathery, broad lobed, rigid, sharp-spined, glaucous, bluish-green, holly-like leaves, with whitish conspicuous edges, nerves and veins,

Fig. 115 Sea Holly *Eryngium maritimum*

and prickly, nearly globular, pale blue or lavender-blue, dense, terminal flower heads. A garden variety is occasionally cultivated. Perennial. Flowering period July and August.

SEA KALE (*Crambe maritima*) [*Cabbage Family—Cruciferae*]
Uncommon. It occurs on sandy and shingle shores, occasionally on sea cliffs. The stems, up to 2 feet in height, bear thick, fleshy, broad oblong, oval-triangular or heart-shaped, glaucous grey-green leaves with the edges lobed, wavy and folded. At the base of the plant the lower leaves are very much larger and thicker, up to 2 feet in length, and prostrate on the ground. The white flowers grow in many-branched wide loose clusters. This plant is the forerunner of our vegetable kale. Formerly its shoots were much gathered as a delicacy. Perennial. Flowering period May to August.

SEA-LAVENDER (*Limonium vulgare*) [*Sea-Lavender Family—Plumbaginaceae*] Fig. 113 (c)
Locally common. It occurs in salt marshes, and on mudflat islands and shores, also sand dunes. The long-stalked broad lance-shaped leaves, with a prominent midrib, form a rosette around the base of the leafless stems bearing the small, crowded, flattish-topped, branched heads of bluish-purple or deep lilac flowers, the wiry flower stems being up to 18 inches in height. The only similarity to the garden lavender is the colour of their flowers. Where widespread in a habitat its flowers create a mass of brightening colour. Perennial. Flowering period July to November. There are related species. The locally common ROCK OR UPRIGHT SEA-LAVENDER (*Limonium binervosum*) occurs on rocky coasts, cliff faces, sometimes on shingle, but is smaller, up to 1: inches in height, with smaller narrower leaves which have up to three veins instead of the single midrib, while the flowers are blue-purple and have white calyx lobes. The rare MATTED SEA-LAVENDER (*Limonium bellidifolium*) is only up to 6 inches in height, the rootstock being branched and matted, the ribbed and three-veined leaves are small and spoon-shaped, and the flowers pinkish-blue or pale lilac. It occurs only in salt marshes in eastern England, chiefly Norfolk. Perennial. Flowering period July and August.

SEA MILKWORT (*Glaux maritima*) [*Primrose Family—Primulaceae*]
Common. It occurs in salt marshes and muddy, sandy estuaries, on rocky coasts and cliffs, sometimes in the splash zone. A small plant, i

stem up to 5 inches in height, bearing small, smooth, thick, fleshy, stalk-less, oval or oblong-narrow leaves, glabrous bluish-green in colour, the upper surface being darker than the underside. The tiny, petal-less, solitary, pale pink flowers appear at the base of the leaves. Perennial. Flowing period May to July.

SEA PANSY (*Viola curtisii*) [*Violet Family—Violaceae*]
Uncommon. It occurs on coastal sand dunes in north and west England, is rare elsewhere. The plant is smaller than the garden pansy and bears yellow flowers only. Flowering period April to August.

SEA PEA (*Lathyrus japonicus*) [*Pea Family—Papilionaceae*]
Uncommon. It occurs on stable shingle and pebbly stony shores in eastern and southern England. The spreading prostrate angular stem, up to 3 feet in length, is secured by short tendrils clasping the stones. Its leaves comprise three to eight pairs of smooth, oval or egg-shaped, pointed, glaucous bluish-green leaflets, which end in a tendril. The typical large pea-like flowers, up to eight on each flower-stalk, are bluish-purple becoming blue as they fade. Perennial. Flowering period June to August.

SEA PEARLWORT (*Sagina maritima*) [*Pink Family—Caryophyllaceae*]
Locally common. It occurs in dry waste places near the sea. It is a very small plant, up to 3 inches in height, erect or prostrate, with wiry stems, forming low tufts, that bear thick, fleshy, blunt, opposite, linear leaves and tiny greenish flowers borne on long stalks. Annual. Flowering period June to September.

SEA PLANTAIN (*Plantago maritima*) [*Plantain Family—Plantaginaceae*] Fig. 113 (e)
Common. It occurs in salt marshes, on cliff bases and ledges, rocky coasts, mudflats, estuary shore edges, seawalls, meadows and banks near the sea. The long flat succulent leaves, up to 12 inches in length, are pointed, narrow linear and have up to five obscure "nerves", but some of them may also be short, very thin and grasslike. The tiny brown and yellow flowers appear together in a dense slender spike, the flower stem being up to 12 inches in height. See BUCK'SHORN PLANTAIN. Perennial. Flowering period June to September.

SEA PURSLANE (*Halimione portulacoides*) [*Goosefoot Family—Chenopodiaceae*] Fig. 116

Common. It occurs on muddy and sandy seashores, mudflat and creek edges, also salt marshes. It is a shrubby, branched, erect and prostrate plant, the upright branches being 18 inches in height, but the straggling leaning or interweaving stems may be longer. The brownish stems bear oval or spoon-shaped, mostly opposite, thick, succulent, blunt or round-tipped leaves, which taper narrowly to the stem. At the ends of the branches are the short dense clusters of tiny yellow-green flowers. The plant is usually immediately noticeable because of the mealy coating of the leaves and stems which give it a conspicuous whiteness in the habitat. Perennial. Flowering period July to October.

Fig. 116 Sea Purslane *Halimione portulacoides*

SEA RADISH (*Raphanus maritimus*) [*Cabbage Family—Crucifera*

Uncommon. It occurs on sea cliffs and seashores, occasionally in sout west England, is common in the Channel Islands, but otherwise scar in Britain. The stems, up to 3 feet in height, bear large coarse pinnatel divided or lobed leaves, comprising alternate large and small lobes, t largest lobe forming the terminal end of the leaf. The veined yellow flowers are on long stalks in a loose flower head. Perennial. Flowering period June to August.

SEA ROCKET (*Cakile maritima*) [*Cabbage Family—Cruciferae*]
Locally common. It occurs on sandy coast and salt marshes near high-tide level and above. It is an untidy bushy glaucous plant, with long straggling numerous zig-zag branches, up to 18 inches in length, bearing smooth, alternate, fleshy, thick leaves deeply cut into several blunt narrow lobes. The clusters of veiny lilac flowers appear at the terminal ends of the branches. Annual. Flowering period June to August.

SEA SANDWORT (*Honkenya peploides*) [*Pink Family—Caryophyllaceae*]
Common. It occurs on stable shingle, pebbly and sandy shores. It is a creeping plant which sometimes carpets large areas of habitat. The prostrate stems bear forked branches, up to 8 inches in height, of numerous, broad oval, pointed, stalkless, fleshy, green leaves and small, five-petalled, greenish-white flowers. Several former botanist-writers confused this plant and called it SEA PURSLANE, the latter really belonging to the Goosefoot Family. Perennial. Flowering period May to August.

SEA SPURGE (*Euphorbia paralias*) [*Spurge Family—Euphorbiaceae*]
Locally common. It occurs on sandy coasts and shores. The erect stout stem, up to 12 inches in height, bears a large number of short, thick, leathery, broad oval, concave, pale green or glaucous greyish leaves close together in whorls up the stem, giving it a bushy appearance. The yellowish-green flowers grow in umbel-like heads at and near the summit of the stem. The plant is poisonous and exudes a milky juice if severed. Perennial. Flowering period from June to October. The locally common, related PORTLAND SPURGE (*Euphorbia portlandica*) occurs on coastal sand dunes, sandy stony coasts and cliffs in south and west England and south-west Scotland. It is smaller, up to 9 inches in height, stem and leaves have a reddish hue, the numerous leaves being thinner, smaller, narrow lance-shaped or oval, greyish, the flower heads being wider apart. Also poisonous. Biennial. Flowering period April to September.

SEA SPURREY OR SHORE SAND-SPURREY (*Spergularia marina*) [*Pink Family—Caryophyllaceae*]

Common. It occurs in salt marshes and muddy brackish areas near the sea. The plant has prostrate, fleshy, branched stems bearing thick, fleshy, narrow, opposite, yellow-green leaves; the five-petalled flowers are pink, occasionally with white centres. Annual. Flowering period June to September. The common GREATER SEA SPURREY OR SALT-MARSH SAND-SPURREY (*Spergularia media*) occurs in muddy salt marshes and similar habitats near the sea. It is a larger more spreading plant than SEA SPURREY, with branched flat-sided stems bearing fleshy blunt-tipped leaves, the larger flowers being lilac pale violet-pink, or almost white when they have white centres Perennial. Flowering period June to September. The locally common CLIFF SPURREY OR ROCK SAND-SPURREY (*Spergularia rupicola*) occurs on rocky shingly habitats and on cliffs, near the sea. It is similar but has hairy reddish-purple stems, the flowers being pink only Perennial. Flowering period June to September.

SEA STORK'S-BILL (*Erodium maritimum*) [*Geranium Family—Geraniaceae*]

Locally common. It occurs in sand dunes, dry stony, grassy, bare, waste places near the sea. The small, softly hairy, viscid, prostrate plant, up to 6 inches in height, bears small, oval or heart-shaped, toothed leaves and pale pink or reddish-purple flowers. The fruit is like a stork's bill hence its name, with a spiral twist. Annual. Flowering period May to September.

SEA WORMWOOD (*Artemisia maritima*) [*Daisy Family—Compositae*]

Locally common. It occurs on salt marshes, seawalls and dry banks near the sea. The branched stems, up to 18 inches in height, bear twice pinnate leaves with narrow linear leaflets. The leaves on each side and the stems are covered with silvery down. The small narrow heads of flowers, in erect or drooping spikes, are reddish or orange. The plant has a rather pungent aroma, but not so strongly as Common Wormwood. Perennial. Flowering period July to September.

SEASIDE THISTLE OR SLENDER-FLOWERED THISTLE (*Carduus tenuiflorus*) [*Daisy Family—Compositae*]

Locally common. It occurs in sandy, waste places, hedges, near the sea. The branched, stiff, slender, downy white stem, up to 4 feet in height has continuous, sharp bristled, toothed "wings", the stalkless leaves, white woolly on their underside, having spined, broad, toothed, wavy lobes. The small pale pinkish-purple flower-heads, in terminal clusters, are oblong and narrow, hence one of its names. Annual or biennial. Flowering period June to August.

TAMARISK (*Tamarix gallica*) [*Tamarisk Family—Tamaricaceae*] Fig. 117

Locally common. It occurs in hedges or independently near the sea, is also frequently planted as a protective windbreak cover for less hardy plants, as it can endure salty spray-carrying winds, also to stabilize sand and loose soils; chiefly in the east, south and west of England. It is a beautiful small tree or shrub, up to 12 feet in height, the delicate, purplish or reddish twigs of the long flexible feathery branches bearing the numerous, tiny, pointed, pale green, scalelike, overlapping leaves. The white or rose-pink flowers are equally minute, crowded in a spike

Fig. 117 Tamarisk *Tamarix gallica*

up to 1½ inches in length. Evergreen. Flowering period July to November.

TASSEL PONDWEED (*Ruppia maritima*) [*Tassel Pondweed Family—Ruppiaceae*]
Locally common. It occurs in salt marsh pools, brackish pools and salt-water ditches. The much-branched weak slender stem, up to 12 inches in length, bears very narrow dark green leaves which have a sheath at their base, from which grow the long stalks of greenish flowers that arise above the surface. This and the following, with EEL GRASS, are some of the few flowering plants which endure being submerged. Perennial. Flowering period July to September. The uncommon COILED PONDWEED (*Ruppia spiralis*), up to 12 inches in length, has wider, dark green leaves, the flower stalks, after maturing, coiling so that the fruits are submerged in the water. It also occurs in brackish and salt marsh pools.

THRIFT OR SEA PINK (*Armeria maritima*) [*Sea-Lavender Family—Plumbaginaceae*] Fig. 113 (f)
Common. It occurs on rocky, muddy, sandy seashores, cliff ledges and summits and is also frequently cultivated in gardens as a rockery or path-edging plant. The tufts of narrow linear grasslike leaves grow together in a habitat to form an extensive cushionlike covering. From the tufts arise the smooth leafless stalks up to 9 inches in height, bearing at their summit the dense, roundish or half-round head of rosy pink to almost white flowers, below which are some thin dry brown bracts which form a sheath round the top of the stalk. The plant is almost unmistakable and was used as an emblem on the reverse of 12-sided threepenny "bits" during the reign of George VI. Perennial. Flowering period April to October.

VIPER'S BUGLOSS (*Echium vulgare*) [*Borage Family—Boraginaceae*]
Common. It occurs on gravelly dunes and dry waste places near the sea, on limestone and chalk cliff tops and ledges, also inland. The erect stem, up to 2½ feet in height, bears numerous rough narrow lance-shaped leaves. The flower buds droop in a dense pink cluster at first, then become reddish-purple, finally brilliant blue, curving, trumpet-shaped flowers from which the pink-purple stamens project. All the plant is covered with stiff, bristly hairs. When a large number of plants are colonising a habitat it is an attractively colourful sight. It is at its

best when the blue flowers and pink buds are on the same plant. Annual or biennial. Flowering period June to September.

WALL PEPPER OR YELLOW BITING STONECROP (*Sedum acre*) [*Stonecrop Family—Crassulaceae*]
Common. It occurs in tufts on stable shingle and sandy dunes, also rocks, wall tops and building rooves near the sea, also inland. The trailing creeping stems, up to 4 inches in height, bear numerous small, thick, oval, glabrous, stalkless, succulent, green leaves. At the summit of the erect leafy flowering stems there is a crowded cluster of golden-yellow, pointed petal, starlike flowers, the other shorter, densely crowded, leafy stems being barren. Where it establishes itself in a suitable habitat the numerous stems creep in all directions and form covering green and yellow mats. If chewed the leaves and stems have a bitter, peppery taste, hence its name. Perennial. Flowering period June and July.

WILD OR SEA CABBAGE (*Brassica oleracea*) [*Cabbage Family—Cruciferae*] Fig. 114 (a)
Rare. It occurs on sea cliffs, sometimes waste places, close to the sea, chiefly in south-west England and Wales. The base of the branched stems, up to 2 feet in height, is thick, tough, woody, sometimes twisted, the stem bearing two types of smooth, fleshy, glaucous, greyish-green leaves. The lower leaves are large, broad, stalked, wavy, irregularly lobed, cabbage-like, up to 18 inches in length; the smaller stalkless upper leaves are oblong. The large lemon-yellow flowers, each about one inch across, form a loose flower spike. This wild plant is the parent of our garden vegetable cabbage varieties and was formerly much gathered. See SEA KALE. Perennial or biennial. Flowering period May to August.

WILD CELERY (*Apium graveolens*) [*Carrot Family—Umbelliferae*]
Locally common. It occurs in salt marshes, ditches and marshy, wet habitats near the sea. The branched leafy stem, up to 2 feet in height, is stout and furrowed. From this plant the Garden Celery was developed, the former having the similar, familiar leaves – large, smooth, shiny, pinnate or divided into three or five lobed, toothed, leaflets. The lower leaves are stalked while the upper smaller trefoil leaflets are stalkess. The tiny greenish-white flowers are grouped together in small compound umbels. An aid to identification from other species is its strong aroma of celery, but the wild celery should *not* be eaten as it has an acrid flavour and is poisonous. Biennial. Flowering period June to August.

Fig. 118 Yellow Horned Poppy *Glaucium flavum*

YELLOW HORNED POPPY OR SEA POPPY (*Glaucium flavum*)
[*Poppy Family—Papaveraceae*] Fig. 118

Locally common. It occurs on sandy dunes and sea coasts, stable shingle and ledges of cliff faces, especially chalk. It is very conspicuous and easy to identify. The branching stems, up to 2 feet in height, bear large thick, leathery, wavy-lobed and cut, glaucous, blue-green or greyish green leaves, which clasp the stems. The four-petalled, golden-yellow sometimes lemon-yellow, flowers are large, up to 3 inches in width. The two-valved seed pods or "horns" are curved and measure up to 12 inches in length. If severed the stems exude a yellow juice and it should be remembered that this seashore beauty is poisonous. Annual or biennial. Flowering period June to September.

YELLOW-WORT (*Blackstonia perfoliata*) [*Gentian Family—Gentianaceae*]
Locally common. It occurs on the grassy banks and turf of chalk and limestone cliff tops and in stable sandy shingle and dunes, also inland. The smooth erect stem, up to 12 inches in height, bears opposite pairs of broad oval pointed leaves which are joined at their base, perfoliate, to form a "collar" which surrounds the stem. At the base of the stem the leaves are spoon-shaped and form a rosette around it. The stem branches towards its summit and these bear the clusters of several bright yellow flowers which open only during sunshine. An aid to identification is that the entire plant has a glaucous greyish-green waxy hue and the yellow parts of the buds have a spiral twist. Annual. Flowering period June to September.

GRASSES

There are just over 150 species of the Grass Family (*Gramineae*) to be found in the British Isles. The largest number are inland grasses, but some of these also occur on or close to the coast, estuary and sea-shore. There are also a lesser number of grasses which thrive in the latter habitats and could not do so elsewhere, the salt content of the soil or mud, a certain degree of aridness or alternatively a frequent washing by spray or near-submergence by the sea, being essential to their existence. Because of these qualities some of the grasses have proved immensely valuable in the fight against erosion and for land reclamation.

Grass identification would seem to be a difficult task for the visitor to coastal habitats and indeed I admit that a specialist's knowledge and microscopic examination is a part of correctly naming some of the inland species which are closely similar. However, those I have described should present no difficulty in recognition.

The grasses are hollow-stemmed, whereas the stems of the petal-less and sepal-less Sedges are never hollow, while Rushes, which appear grasslike, differ by having normally structured flowers, with petals, sepals and stamens, and may have pith-filled stems. Grass flowers do not have petals or sepals and are usually very small, with stamens, stigmas and ovary. The tiny flowers are contained in leaf-like scales, to form spikelets, sometimes with other leaf-like scales. These spikelets compose the varying forms of grass flower-heads, depending on how they are made up, some, for example, being densely crowded spikes while others are on loose, delicately hanging branches. What is familiarly called the "flower" of a grass plant is in fact a number of the tiny flowers largely concealed in the spikelet's scales. Some of the grasses produce a large amount of pollen, as hay fever sufferers well know, the grass flowers usually being fertilised through the action of the wind. Grasses spread by seed, or vegetatively, either by a creeping rootstock thrusting up new shoots a distance from the parent, or by extending their stems over the soil surface.

COMMON CORD-GRASS OR RICE GRASS (*Spartina townsendii*) Fig. 119 (a)
Locally common. Occurs on estuary, mudflat edges and soft coastal mud, sometimes as large, circular, extending tufts. The tough, stout, erect flower stem, up to 4 feet in height, bears thick, stiff, ribbed inrolled or flat, yellow-green or bright green leaves which taper to a long, hard point. The flower-head comprises several erect or spreading, unbranched, dense, stiff spikes of overlapping, tightly pressed, yellowish-brown spikelets. It was first discovered on the Hampshire coast in 1870 and since then has spread rapidly along the south coast and extended its range elsewhere, continuing to do so. The latter is partly due to it also being deliberately planted on vulnerable sea coasts for land reclamation because its creeping system of deep roots and rhizomes stabilizes the mud while the culms or stems and leaves trap silt which eventually heightens the mudflat's level and creates salt marshes and meadows. Perennial. Flowering period July to November.

MARRAM GRASS, SEA MATGRASS OR SEA-REED (*Ammophila arenaria*) Fig. 119 (b)
Locally common. Occurs on sand dunes and sandy seashores. The stiff, erect culm or stem, up to 4 feet in height, bears very long, broad, rigid, pointed, grey-green or sea green leaves, with their edges inrolled. The flower-head is a narrow oblong, cylindrical, tapering, whitish-green spike, with yellow anthers, up to five inches in length, of crowded spikelets. It frequently occurs in large spreading clumps and is widely planted as a stabilizer of sand, see SAND DUNES in Chapter 1. Perennial. Flowering period from June to August.

SAND COUCH GRASS (*Agropyron junceiforme*) Fig. 119 (c)
Common. Occurs on sandy seashores and in coastal dunes much closer

Fig. 119 (a) Common Cord Grass or Rice Grass *Spartina townsendii*
(b) Marram Grass *Ammophila arenaria*
(c) Sand Couch Grass *Agropyron junceiforme*
(d) Sea Lyme Grass *Elymus arenarius*

to the sea than other grass species. The stem, up to 3 feet in height, is either erect, spreading or drooping, bearing flat or inrolled, semi-erect or drooping, pointed, ribbed, bluish-grey leaves. The curving or straight flower spikes comprise stalkless spikelets alternate on each side of the stem. This grass, like couch grass which is an agricultural and garden pest, spreads by means of its long wiry rhizomes to form extensive tufts in the habitat. Perennial. Flowering period June to August.

SEA LYME GRASS (*Elymus arenarius*) Fig. 119 (d)
Locally common. It occurs on sand dunes. The stout, unbranched, erect or spreading stem or culm, up to 4 feet in height, bears broad, rigid, flat or edges unrolled, finely pointed, bluish-green leaves. The large glaucous flower spikes, up to 5 inches in length, comprise numerous stalkless spikelets in alternate pairs on each side of the stem. The creeping rhizomes also make this a valuable grass for stabilizing sand, for which purpose it has been extensively planted, sometimes among or close to marram grass. Perennial. Flowering period June to August.

Chapter 17

Insects

The Insects (*Arthropods*) chiefly occur on land, but a small number have extended on to the shore and colonised it, some even adapting themselves to being immersed periodically by the sea. There are also numbers of common insects, wasps, various bees, butterflies, various flies, beetles, etc., which wander on to the coast, estuary or seashore from inland, more so where gardens, marshes, grazing land or cliff tops with abundant plant life are in close proximity to the shore, but these latter insects do not come within the scope of this book. Those inland insects which migrate to this country are, however, dealt with in Chapter 18.

The majority of the insects that can actually thrive on the shore are small, even minute, in size and have to be keenly searched for, although when they occur in large numbers some species become pests and may make their undesired presence known! They are usually found on or among rotting seaweed and other debris on the strand-line or higher up the shore.

BEETLES (*Coleoptera*)

BLEDIUS DIOTA

Fairly common. It frequents the upper shore of coasts and estuaries, where it occurs on and in sandy areas, also in muddy salt-marsh creeks and inlets. A clue to its presence are small heaps of damp sand or mud, similar to worm coils, in which it burrows, although it feeds on seaweed. The beetle is black, with red or brownish-red wing-cases and measures up to $\frac{1}{8}$ inch in length.

DYSCHIRIUS ARENOSUS

Fairly common. It frequents the same habitats as BLEDIUS DIOTA on which it is parasitic, burrowing into the sand to pursue its prey in the victim's burrows. It is brown in colour and measures up to $\frac{1}{8}$ inch in length.

HETEROCERUS FLEXUOSUS
Common. It frequents muddy creeks of estuaries and salt marshes, also similar coastal habitats, where it burrows to create galleries in the mud. Sometimes the mere tread of the human foot nearby will cause it to emerge. It is blackish with brown mottling on the wing cases and measures up to $\frac{1}{8}$ inch in length.

SHORE SEXTON (*Necrodes littoralis*) Fig. 120
Fairly common. It occurs in or under the bodies of dead sea birds and fish, or under stones, on the shore, sometimes in the process of burying the corpse so that it will decompose and provide a habitat for the beetle's larvae. It is black in colour, the club-like tips of the antennae or "feelers" being red, and measures up to 1 inch in length.

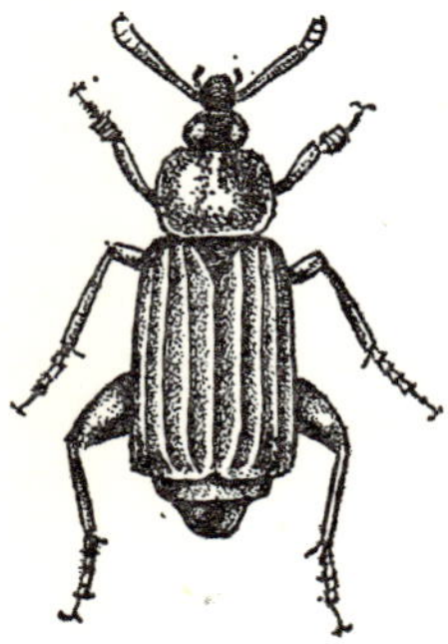

Fig. 120 Shore Sexton (male) *Necrodes littoralis*

Several other very small beetles – CAFIUS XANTHALOMA, DIGLOTTA MERSA, AEPUS MARINUS, MICRALYMMA MARINUM – occur under cast-up seaweed and other debris, in rock crevices, or hiding under stones.

FLIES – (*Diptera*)

FUCELLIA MARITIMA Fig. 121
Occasionally very common. It frequents seaweed that has been cast

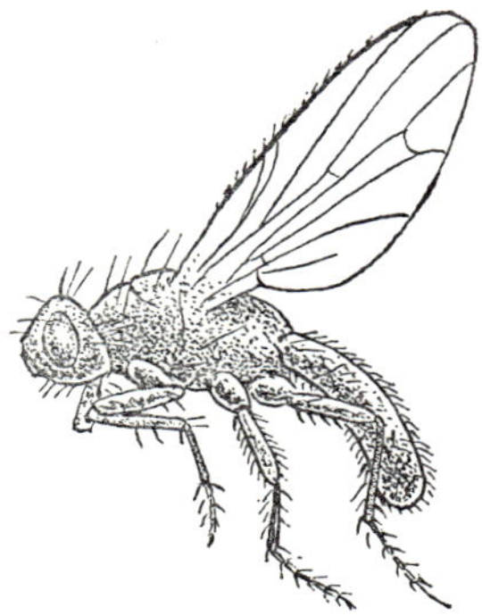

Fig. 121 Fucellia maritima

ashore and is decaying, where it lays its eggs. It is two-winged and has a blackish-grey body, measuring $\frac{1}{4}$ inch in length.

SEAWEED FLY (*Coelopa frigida*)
Occasionally very common. Similar in habits to above, wherever there are large amounts of rotting seaweed, wherein it lays its eggs and has to be dealt with when it increases so much that it becomes a pest. It is dark brown or blackish in colour, two-winged and measures $\frac{1}{8}$th inch in length.

Minute MIDGES in this group also occur on the coast, estuary and sea shore. Two of the commonest are THALASSOMYIA FRAUENFELDI and CLUNIO MARINUS. The larvae of the latter lives upon seaweed, while the male adult rarely flies and the female is wingless.

MISCELLANEOUS

BRISTLE TAIL OR BRONZE-FISH (*Petrobius maritimus*)
[*Thysanura*]
Common. It frequents the upper shore, where it occurs hiding in rock crevices and under stones and seaweed, but moves rapidly or leaps into new shelter when revealed. The wingless body is long and narrow, bearing a long "filament", with two shorter "filaments", like a prong, from its tail or hind body. On the head it has two antennae which are the length or longer than the body. It is brownish-grey or metallic brown with darker markings and measures up to $\frac{1}{2}$ inch in length.

SPRING TAIL (*Anurida maritima*) [*Collembola*]
Common. It frequents the upper shore where it occurs on sand, rocks and seaweed, also on the surface of small rock pools, often in large numbers. This is one of the few marine insects. It can survive, by hiding in rock crevices and sand, being submerged for short periods by the sea. It is blue-black and measures $\frac{1}{8}$ inch in length.

Chapter 18

Bird, Insect and Plant Migration

The flora and fauna population of the British Isles is not static, but continuously being added to by new species. The resident native species' numbers are also supplemented by arrivals of the same species from overseas. Being an island these immigrant species have to cross the sea to reach Britain and it is on the coast, seashore and in estuaries that these species can be seen, sometimes in huge numbers.

Perhaps the most familiar is the migration of birds, such as Swallows, House Martins, Cuckoos, etc., which are welcomed as summer visitors. The reason why and how birds migrate here is too complex to be dealt with in this chapter, but basically it is a periodic movement between two habitats. For a summer visitor Britain is a cool habitat for breeding and a warm habitat in the non-breeding season, but for winter visitors, Fieldfares, Redwings, Bramblings, species of Duck, Waders, etc., Britain is a warm winter habitat for their non-breeding season. The largest numbers of certain species arrive on the south, south-east and south-west coast, being the closest areas to the Continent, before dispersing inland. The east and north-east coast receive species from the Continent and Scandinavia, while the north and west receive species migrating from Scandinavia and the Arctic regions and in the appropriate season the species return via these areas. From July onwards into autumn, depending on the weather, Wheatears, Swallows, House Martins, Warblers, etc. can be seen arriving on the coast and perhaps resting there before setting off on their journey back across the sea. The distances covered by some species is remarkable. Swallows, for example, have covered 2,000 miles in approximately 100 days from Africa, equalled by Willow Warblers' 2,500 mile journey in approximately 80 days. Just as marvellous a feat is the crossing of the North Sea from the Continent by Firecrests and Goldcrests, Britain's smallest birds, measuring between 3 and 4 inches in length. In autumn large numbers of Starlings and various Finches, also lesser flocks of Robins, Tits, Pigeons, Skylarks, Wagtails and other species cross the east coast en route from the Continent, some of these having come from as far afield as Russia and East Germany. Knowledge of birds' migratory movements has been obtained after years of research by the British Trust for Ornithology, which operates a bird marking scheme. Birds are harmlessly trapped at observer stations and an aluminium ring, of varying size depending on the species, stamped with a number and address is fixed to the leg of each bird. The species, age and sex of the bird, date and registered number of the station is recorded, then the bird is released. If the bird is shot or killed in any other way, or re-trapped, the ring is removed and returned to the address on it, where details of the bird's progress will then be known. In this way Teal "ringed" in Pembroke have been trapped only a few months later as far away as Russia and Finland. Wigeon "ringed" in England have turned up near the Caspian Sea. The Royal Society for the Protection of Birds, Nature Trusts and other similar organisations have reserves and observatories at various places, one being at Dungeness, a shingle area in south-east Kent, where the trapping and ringing of birds and their migration can be seen, sometimes while sheltered in "hides", but permission must be obtained from the Warden beforehand and it is obviously an advantage to be a member of the relevant organisation. Other good observation points are Blakeney Point, Scolt Head, Holy Island near the Farne Islands, Spurn Head and Fair Isle.

For many years it has been known that insects migrate to and from the Continent over our coasts and seashore. Ships far out at sea have recorded varying numbers of insects, not being carried aimlessly by the wind, but progressing in a deliberate direction. Lightships have also proved valuable in observing insects at sea, especially at night. Ladybird beetles have arrived on the south coast apparently from France. In 1952 a huge swarm of these beetles crossed the Lincolnshire coast and moved inland. Dragonflies have been seen migrating from England and rare species have migrated here in considerable numbers from the Continent. On one occasion I was standing on a lower shore in Kent when I noticed that my clothes were crawling with aphids (greenflies) which were arriving from an easterly direction. As there was only sea in front of me for many miles I could only assume that these were being carried on wind currents from the Continent, and although not a migration in the normal sense it showed how insects move from one country to another despite a water barrier. On another occasion I was cycling up a hill in Kent at night when I heard a thud against my front lamp. Dismounting I shone the lamp on the ground and discovered a stunned Convolvulus Hawk, a moth with a 4½ inch wing span, that arrives from the Continent fairly regularly every year. Numerous other moths, some small in size, such as the Silver Y, Humming Bird Hawk, Angleshades, Dark Swordgrass, The Ni, to name a few, are known to migrate here from Europe.

Butterflies also achieve this. In some instances rare butterflies have suddenly for one summer become common. In 1945 the Bath White, a rare irregular wanderer from Spain and North Africa, arrived in large numbers. Startled lepidopterists captured and recorded 650 throughout Britain but no doubt many more were undetected. In 1946 only five were seen.

The Clouded Yellow, which lives around the Mediterranean coast, sometimes also arrives in large numbers in southern England. In October, 1947, a naturalist aboard a steamer in the English Channel passed a huge swarm, which he estimated contained around 100,000 examples, flying towards the English south coast. Other migratory butterflies are the Painted Lady, Camberwell Beauty, Queen of Spain Fritillary, Pale Clouded Yellow, Red Admiral, Large, Small and Green-veined Whites, the Small Tortoiseshell and Peacock. In 1943 an estimated 15 million "White" butterflies arrived from northern Europe and ravaged crops in Scotland and the Hebrides.

The Milkweed or Monarch butterfly of North America, which cannot breed in Europe because its food plant, milkweed, does not occur here, has been recorded 200 times in Britain since 1876 when it was first seen. These butterflies migrate over 2,000 miles in North America and some of them somehow cross the Atlantic. The most obvious theory is that they stow away aboard ships, but on one occasion a steamer met several 400 miles from land flying towards Britain. The most likely method is that they are wind-borne across the North Atlantic and able to survive the vast distance, but do eventually die here.

Perhaps even more remarkable is that the descendants of some of these migratory butterflies, such as Large White, Small White, Painted Lady and Clouded Yellow, during autumn no doubt try to return to their Continental habitats. The observer may be standing on a shore when a butterfly approaches from the land, flies backwards and forwards along the waves, then sets off, sometimes perilously low above the water, in the direction of Europe and it is known some do arrive back there safely.

Butterflies are also marked to study their migrations and records of these kept. The wings may bear a registration number and letter, stamped or carefully written on, or a printed label with minute address lettering. Other methods used have been aniline dyes to conspicuously stain the wings, or feeding the butterfly's caterpillars with a material that causes the resulting adult butterflies to have an obvious colour change in their wings.

Unlikely as it may seem, plants also migrate via our coasts and shores. Large numbers of seeds are carried from the New World by the Gulf Stream and North Atlantic Drift, but very few are known to germinate on our shores, perhaps because the plants or shrubs cannot thrive in this salt-laden soil or the climate is too cold. The most probable habitats for these alien plants are in ports or their vicinity, after the seeds or roots have accidentally been transported here from overseas, in wool, grain, bird seed, hides and with other commodities. Here they may survive for a season or two, being eventually killed by climate, because the resulting seeds do not ripen, or losing the fight with established British plants. Some thrive, however. In 1809 troops returning from Holland dumped fodder for their horses where they landed in Thanet, Kent. Among this were seeds of the Thanetweed or Hoary-Cress. Since then this has become a pest on the Kent coast, spread inland and to other counties. Another example is Red Spur-Valerian, familiar on cliffs and walls near the sea. This originally came from southern Europe and the Lebanon. How it arrived is unknown, perhaps it was brought back originally as a garden plant but there it is now on many a seashore and coast habitat.

In the summer of 1956 I was walking near the Fal Estuary in Cornwall when I noticed an area of a daisy-like flower. It puzzled me so I posthaste sent some to a botanist friend, the late George Coxhead. He identified it as a species called the Mexican Fleabane. How did it come to be growing in Cornwall? Who knows?

Suggested Books For Further Reading To Increase Knowledge of This Subject:

"The Sea Shore", by C. M. Yonge (Collins)
"Life on the Seashore", by A. J. Southwood (Heinemann)
"Seashore Life In Colour", by G. Vevers (Blandford)
"Pocket Guide to the Sea Shore", by J. Barrett & C. M. Yonge (Collins)
"The Sea Coast", by J. A. Steers (Collins)
"The Shell Bird Book", by J. Fisher (Ebury Press & Michael Joseph)
"A Field Guide to the Birds of Great Britain & Europe", by R. Petersen, G. Mountfort & P. A. D. Hollom (Collins)
"Sea Birds", by D. Sanders (Hamlyn)
"Pocket Guide to Wild Flowers", by D. McClintock & R. S. R. Fitter (Collins)
"Grasses", by C. E. Hubbard (Penguin)
"Observer's Book of British Grasses, Sedges & Rushes", by W. J. Stokoe (Warne)
"The Amateur Geologist", by P. Cattermole (Lutterworth)
"British Shells", by N. F. McMillan (Warne)
"Insect Migration", by C. B. Williams (Collins)

Index